MICROSCOPY HANDBOOKS 14

Enzyme histochemistry

J.D. Bancroft and N.M. Hand

Histopathology Department
University Hospital
Queen's Medical Centre
Nottingham

Oxford University Press · Royal Microscopical Society · 1987

Oxford University Press, Walton Street, Oxford OX2 6DP

Oxford New York Toronto
Delhi Bombay Calcutta Madras Karachi
Petaling Jaya Singapore Hong Kong Tokyo
Nairobi Dar es Salaam Cape Town
Melbourne Auckland

and associated companies in
Beirut Berlin Ibadan Nicosia

Oxford is a trade mark of Oxford University Press

Royal Microscopical Society,
37/38 St. Clements,
Oxford OX4 1AJ

Published in the United States
by Oxford University Press, New York

British Library Cataloguing in Publication Data
Bancroft, John D.
Enzyme histochemistry. — (Microscopy
handbooks; v. 14).
1. Enzymes   2. Histochemistry
I. Title   II. Hand, N.M.   III. Royal
Microscopical Society   IV. Series
574.19'25   QP601
ISBN 0-19-856416-3

Library of Congress Cataloging in Publication Data
Bancroft, John D.
Enzyme histochemistry.
(Microscopy handbooks ; 14)
1. Enzymes—Analysis—Handbooks, manuals, etc.
2. Histochemistry—Handbooks, manuals, etc.   I. Hand,
N.M.   II. Title.   III. Series.   [DNLM: 1. Enzymes.
2. Histochemistry. QU 135 B213e]
RB48,B35   1987   616.07'583   87-5588
ISBN 0-19-856416-3

Typeset by Grestun Graphics, Abingdon, Oxon
Printed in Great Britain by
The Alden Press, Oxford

Contents

Acknowledgements

We are grateful to Professor D.R. Turner for his encouragement and understanding in the production of this handbook. Our thanks also go to Miss Jane Marsh who typed the manuscript.

For Carol and Jacqueline

1

Introduction

1.1. Definitions

Enzymes are vital components of biological systems; as the catalysts of most biochemical reactions they are essential for the metabolic processes occurring within tissues.

Catalysts lower the energy of activation, bringing molecules into a reactive state without themselves being consumed in the process. It is a characteristic of catalysts that there is no stoichiometric relationship to the quantity of substance catalysed; only a small concentration of enzyme is necessary to be effective.

Enzymes are protein molecules of high molecular weight often containing a non-protein prosthetic group. The active protein moiety (the *apoenzyme*) will contain various ionized groups, e.g. carboxyl and amino. Enzymes may be free and soluble in the cytoplasm or body fluids (*lysoenzymes*) or bound to specific cell components (*desmoenzymes*).

To achieve a satisfactory enzymatic reaction the presence of *cofactors* is required. These are often metal ions, frequently magnesium and manganese (activators) and compounds such as the nucleotides – nicotinamide adenine dinucleotide (NAD) and nicotinamide adenine dinucleotide phosphate (NADP), which are *coenzymes*. Enzymes are classified into groups according to effect on substrates (see section 2.1). In practice, most techniques for histochemical demonstration are for hydrolytic and oxidative enzymes.

1.2. The development of enzyme histochemistry

Enzymes have long been assayed biochemically. The first diagnostic application of a serum enzyme assay was made by Kay in 1930 when it was shown that raised levels of the enzyme alkaline phosphatase were found in certain bone diseases and obstructive jaundice. Later, the detection and localization of enzymes were sought in histological sections and, according to Pearse, in 1953 techniques for only 18 enzymes were available. The growth of enzyme histochemistry may be measured by noting that during the late 1960s the number had risen to around 75 and today the figure is in excess of a hundred. In the early 1950s, one of the pioneer enzyme histochemists, Gomori, summarized his work by publishing, in 1952 *Microscopic Histochemistry*. In addition, the first edition of *Histochemistry* by Pearse in 1953 greatly stimulated interest in enzyme histochemistry and today the further editions remain authoritative texts.

Enzyme histochemical methods enable the identification and localization of specific enzymes within tissue. The methods depend on the visual identification of reaction products generated from chemical substrates by the catalytic influences of the enzymes in question. Some enzymes are extremely labile and factors including temperature, pH, and fixation need to be carefully considered as they seriously affect enzyme activity.

Although paraffin wax sections are routinely used in histology they are of less value in enzyme histochemistry as a loss of enzymes occurs during processing. For this reason, enzyme reactions are usually performed on frozen sections. The crucial improvements in enzyme histochemistry emerged after the development of satisfactory methods for generating this type of tissue preparation. In particular the *cryostat*, first developed in 1938, became more sophisticated and reliable during the 1950s.

The intrinsic water content of the tissue, when frozen, acts as a support medium for the tissue and renders it rigid enough for thin sections to be cut. The formation of ice within the tissue and the subsequent dissolution of ice on thawing leads to some architectural distortion and diffusion, but many enzymes in which the histochemist is interested remain and may be demonstrated by histochemical reactions. Sections produced in this way contain the full complement of substances present *in vivo*, and the tissue has been exposed to no damaging agent other than a low temperature, which seems to matter little from a practical point of view. The 1950s also saw improvements of incubation techniques with the exploitation of diazonium and tetrazolium salts, and substituted naphthols, as histochemical reagents.

Enzymes are located in numerous sites; they are found in intracelluar organelles such as the Golgi apparatus, lysosomes and mitochondria or on the brush border of enterocytes and along nerve-fibre tracts. Their localization and demonstration are not only useful morphological markers, but also indicate a cell or structure's function, a characteristic which is of significance in diagnostic histopathology.

The subject of enzymes and their demonstration is too large to be covered comprehensively in this handbook. The purpose of the present work is therefore to introduce the reader to theoretical and practical concepts which may be of help in the laboratory and stimulate further reading. Only the most popular methods, either as a result of their reliability or diagnostic importance, are discussed. For further reading, Bancroft and Stevens (1982), Filipe and Lake (1983), and Pearse (1968, 1972) are recommended.

Classification and terminology

2.1. Introduction

The nomenclature of enzymes was originally haphazard and confusing. On occasion two or more names had been used for the same enzyme by different workers, or conversely the same name had been used for two or more enzymes. In 1898, Duclaux suggested enzymes be named after the substrate upon which they acted with the added suffix -*ase*. For example, the term suc*rase* indicates an enzyme acting on suc*rose*. Later it became necessary to name not only the substrate but also the type of reaction, e.g. cholinesterase. Further definition also proved necessary to differentiate between two enzymes catalysing similar reactions, but under different conditions, e.g. acid phosphatase and alkaline phosphatase.

2.2. Classification

As the number of enzymes identified increased such a simple nomenclature became increasingly inadequate. Several attempts were made to devise an acceptable system of nomenclature and classification, and in 1956 the Commission on Enzymes was set up by the International Union of Biochemistry. Their report which was accepted in 1961, is based on two features. The chemical reaction catalysed is the specific characteristic which distinguishes one enzyme from another. This forms the basis of a classification which segregates enzymes into groups, each group containing enzymes that catalyse similar processes, with sub-groups specifying the reaction more precisely.

Each enzyme is referred to by its chemical substrate name followed by a word with the suffix -*ase* specifying the type of reaction involved, e.g. L-lactate NAD^+ oxidoreductase. This is known as the *systematic name*. This terminology is precise but many names are too cumbersome for everyday use. The Commission suggested a series of *trivial names* for common use, such as lactate dehyrogenase, which is the name for the above enzyme.

The Commission also adopted a scheme for numbering enzymes which is closely allied with the above nomenclature and classification. The enzyme code number (E.C. No.) comprises four figures separated by points. The first figure denotes which of the six classes an enzyme belongs (Table 1). Further sub-division is indicated by the other figures. The second figure is a sub-group of the first figure and indicates the type of group involved in the reaction, e.g. hydrolases may act on esters giving rise to the code 3.1. The third figure is a sub-group of the second and characterizes the reaction more precisely, e.g. a code number of 3.1.1 refers to a hydrolase acting on esters of a carboxylic acid. Finally, to complete the enzyme

Table 1. *Classes of enzymes*

First figure of E.C. No.	Class
1	Oxidoreductases
2	Transferases
3	Hydrolases
4	Lyases
5	Isomerases
6	Ligases

code number, the fourth figure is the serial number of the enzyme. For example, 3.1.1.8 has the systematic name of acylcholine acyl-hydrolase or as it is referred to more commonly, cholinesterase.

Such a numbering scheme also has the desirable feature that when a new enzyme is discovered it fits conveniently into the system without alteration to name and number of other enzymes. A comprehensive classification list and description of the various sub-groups is beyond the scope of this book, but readers who wish for more detailed information should consult the works of Wilkinson (1976) and Dixon and Webb (1979). The following six classes are worthy of brief discussion, however.

Oxidoreductases. These are a large and important group to the enzyme histochemist. These enzymes were previously known as oxidases and dehydrogenases and are often referred to as oxidative enzymes. The following are included in the class oxidoreductases.

Oxidases: Catalyse oxidation of a substrate in the presence of oxygen.
Peroxidases: Catalyse oxidation of a substrate by removing hydrogen which combines with hydrogen peroxide.
Dehydrogenases: Catalyse oxidation of a substrate by removal of hydrogen.
Diaphorases: Catalyse oxidation of NADH and NADPH by removal of hydrogen.

Transferases. These enzymes catalyse the transfer of the radicals of two compounds without the loss or uptake of water. There are several sub-groups of transferases.

Hydrolases. In most cases these enzymes catalyse the introduction of water or its elements into specific substrate bonds, although in some instances water may be removed. Hydrolases include esterases, lipases, phosphatases, glycosidases, peptidases, and pyrophosphatases, and form a very important group in enzyme histochemistry.

Lyases. These catalyse the removal of groups from substrates by mechanisms other than hydrolysis; this process results in the formation of carbon--carbon double bonds. A sub-class is *decarboxylases.*

Isomerases These catalyse molecular rearrangements of a compound.

Ligases. These catalyse the linking of two compounds, coupled to the breaking of a pyrophosphate bond in ATP or a similar compound, with the release of energy.

The above classification of enzymes is biochemical, and in histochemistry the situation is less precise. Although the number of enzymes that may be histochemically demonstrated is at present in excess of one hundred, there are few methods for lyases and none for isomerases and ligases. Methods exist for transferases but most are for hydrolases and oxidoreductases, commonly referred to as *hydrolytic* and *oxidative* enzymes, respectively.

3

Preservation of enzymes

3.1. Introduction

Owing to the labile nature of enzymes, specific procedures are required for their preservation in tissues. Enzymes respond in various ways to outside influences. Mitochondria containing many oxidative enzymes are rapidly injured when denied a supply of oxygen as the membranes become damaged and a loss of enzyme activity occurs. Lysosomes containing hydrolytic enzymes are damaged by freezing and thawing of tissue blocks and sections, which is shown by diffusion of the enzymes. This diffusion may be minimized by fixing the tissue. Most mitochondrial enzymes are less sensitive to the effects of freezing and thawing, as this appears not to rupture the mitochondrial membrane.

Many oxidative enzymes are lost if the tissue is fixed prior to histochemical demonstration, but to minimize diffusion, the section is fixed after incubation. Hydrolytic enzymes are able to withstand some fixation, which improves localization, although there may be some enzyme loss. As a result, enzyme histochemistry for oxidative enzymes is carried out on unfixed sections whereas tissue for the demonstration of hydrolytic enzymes is usually fixed prior to incubation. Although fixation may be critical, there are several other factors which can also influence enzyme preservation and subsequent demonstration.

3.2. Fixation

The adverse effect of fixatives on hydrolytic enzymes prompted the use of alcohol or acetone in early histochemical studies as these two protein precipitants produced only a small loss of enzyme activity. Tissue morphology of frozen sections cut after fixation in these solutions is poor. The introduction of formalin fixation by Seligman *et al.* (1951) improved both the preservation of enzymes and tissue structure. Following this, numerous studies on the effects of aldehyde fixation have been reported. Glutaraldehyde and paraformaldehyde have been used for electron histochemistry, but for light microscopy 4 per cent formol calcium as recommended by Dawson (1972) remains a good fixative, with good preservation of structure and enzyme activity.

3.3. Temperature

Enzymes are rapidly lost if fresh tissue is left at room temperature or higher. Tissue which is required unfixed for the demonstration of oxidative enzymes should be frozen as soon as possible. Fixation for demonstrating hydrolytic enzymes is at 4°C. The fixatives should be chilled before the tissue is immersed.

3.4. Fixation pH

Enzymes can be destroyed if fixed at a pH greatly different from the physiological pH of 7.4. Consequently, fixatives utilized for enzyme histochemistry are around pH 7.0.

3.5. Fixation time

The longer tissue is fixed, the less enzyme activity survives. The rate of loss depends on the enzyme and fixative, but many hydrolytic enzymes are well demonstrated after 18 h in formol calcium or 4 h in glutaraldehyde. Tissue is thinly sliced (1–2 mm), as the rate of penetration of fixative is retarded at 4°C.

3.6. Washing

Several studies have shown that washing the tissue after fixation improves enzyme activity. Sucrose solutions have gained acceptance, with hypertonic gum sucrose (Holt *et al.* 1960) used for frozen sections. This solution also improves frozen tissue morphology. Washing of the tissue at 4°C for several hours (overnight is often convenient) is recommended. Before freezing the tissue, excess gum sucrose is removed by blotting.

4

Specimen preparation

4.1. Introduction

A number of different methods are available for the examination of histological material. Sections for enzyme histochemistry are usually frozen, but several other preparations may also be used. The techniques can be considered under the following headings: *Frozen sections* (4.2); *freeze drying* (4.6); *freeze substitution* (4.7); *paraffin sections* (4.8); *resin sections (for light and electron microscopy)* (4.9); and *smears* (4.10).

4.2. Frozen sections

These are produced by using a freezing *microtome* or *cryostat*. The principle of cutting frozen sections is based on the water in the tissue turning to ice when the tissue is frozen. The consistency of the frozen tissue is altered by varying the temperature; lowering it produces a harder block and by raising it, the tissue becomes softer. A suitable cutting temperature depends upon the nature of the tissue; the majority of non-fatty tissues section well around −20°C. Those containing a large amount of lipid, such as in breast tissue, require a cutting temperature of −30°C.

Sectioning fixed tissue requires a block temperature around −10°C. The increased water content in a fixed block produces a harder consistency and a higher temperature is needed to obtain the ideal consistency for sectioning. The temperature of the tissue block can also be altered during sectioning by using a thermomodule as the tissue block holder on the freezing microtome or cryostat. These utilize the *Peltier effect*, and the temperature of the surface is rapidly altered by controlling the direct electrical current. A crude method of raising the temperature of the block is to pass a finger over the surface of the tissue. This is common practice when using the freezing microtome but is not recommended if enzyme histochemistry is required, as temporary thawing of the surface of the block causes diffusion of tissue constituents.

4.2.1. Methods for freezing tissue

Fresh tissue to be used for the demonstration of oxidative enzymes is rapidly frozen as soon as possible after removal. Slow freezing causes distortion of tissues owing to ice crystal artefact; the more rapid freezing, the smaller the ice crystals formed. A number of freezing agents are available and include:

Liquefied gases: e.g. nitrogen (−190°C)
Isopentane cooled by liquid nitrogen (approx. −150°C)
Solid carbon dioxide: 'cardice' (−70°C)
Carbon dioxide gas: (−70°)
Aerosol sprays: approx. (−50°)

The use of liquified gases is restricted to specialized laboratories. Liquid nitrogen, which is often used, has the advantage of rapid freezing (also termed *quenching* or *snap freezing*), but has a low rate of thermal conductivity; a layer of vaporized gas forms around the specimen, slowing down transfer of heat from the tissue to the freezing medium. Consequently, other coolants, such as isopentane super-cooled with liquid nitrogen, are preferred.

Rapid freezing can cause tissues to crack if the block is large, owing to the contraction of the tissue and the expansion of water (intrinsic or extrinsic) during freezing. The use of isopentane cooled by liquid nitrogen is recommended for enzyme histochemistry but, if not available, isopentane pre-cooled in a cardice-acetone mixture at −70°C is satisfactory. Solid carbon dioxide can also be used to freeze tissue blocks. Two pieces of cardice are held in gloved hands against the cryostat block-holder supporting the tissue. As the tissue is frozen, a white line will be seen to pass through the tissue.

In recent years the use of aerosol sprays containing liquid gas under pressure, e.g. Arcton-12 (dichlorofluromethane), has become increasingly popular. *Care should be exercised when using these as there are reports of lung damage.* The sprays are adequate for freezing small blocks of tissue, with the exception of muscle. These sprays are readily available from a number of suppliers, and are easily stored.

4.2.2. Method for freezing of unfixed tissue

The use of a piece of cork between the block-holder and tissue is recommended if the tissue is small or storage of the specimen is required.
1. Place 50 ml of isopentane in a Pyrex beaker, cool in liquid nitrogen until the isopentane is solid.
2. Remove the isopentane from the nitrogen allowing it to partially thaw so the tissue can be immersed.
3. Place OCT (Tissue Tek) compound or water on the block-holder, or a piece of cork, and orientate the specimen.
4. Immerse the specimen in thawing isopentane until frozen.
5. If the specimen is on cork, freeze the specimen in thawing isopentane. It is attached to the block holder by freezing with OCT compound or water.
6. Place the frozen block into the cryostat.

4.2.3. Preparation of fixed tissue

The use of unfixed tissue causes diffusion of some enzymes. This is increased when the section is cut in the cryostat where the heat produced on sectioning causes a fractional thawing of the cut surface. Thawing also occurs when the section is picked up on a warm slide or cover-slip. This is unavoidable when oxidative enzymes are to be demonstrated; however, they are more resistant to the effects of thawing than hydrolytic enzymes. To obtain accurate localization of some hydrolytic enzymes, it is important to fix the tissue before sectioning. Tissues prepared in this way are fixed under controlled conditions. It has to be fresh and placed in formol calcium at 4°C. The block is fixed for up to 18 h.

METHOD

1. Fix fresh tissue block in formol calcium at 4°C for up to 18 h
2. Rinse in running tap water
3. Blot dry
4. Place tissue in gum sucrose solution (see p. 59) at 4°C for 18 h
5. Blot dry
6. Freeze tissue onto block holder

4.3. Freezing microtome

There are a number of microtomes available which are basically similar in design and operation; the knife passes over the surface of the block, in contrast to paraffin microtomy where the opposite applies. The microtome block-holder is either attached to a supply of carbon dioxide gas or incorporates a thermomodule.

The thermomodule, in addition to freezing the tissue, has the advantage of keeping it at the correct cutting temperature. On freezing microtomes modified by thermomodules, sections from 5 μm are reasonably easy to produce. Using carbon dioxide, sections consistently thinner than 10 μm are difficult to cut, and alternate thick and thin sections are often produced. Thin sections will break up during subsequent handling. For enzyme histochemistry, sections are floated onto either a fixative or distilled water at room temperature and picked up on slides or cover-slips coated in adhesive (see p. 59). The sections are dried at room temperature. This technique gives acceptable results for fixed tissue. The freezing microtome is not recommended for sectioning unfixed tissue for oxidative enzyme histochemistry.

4.4. Cryotomy

A cryostat is a refrigerated cabinet (−5 to −40°C) in which a modified microtome is housed and operated with controls on the outside. To produce good thin sections a number of conditions need to be fulfilled. Tissue has to be adequately prepared by one of the techniques outlined above; the conditions in the cryostat have to be correct. These conditions include:

Correct cabinet temperature
Microtome operating correctly
Anti-roll plate adjustment correct

4.4.1. Tissue block temperature

When the tissue block is ready for sectioning, the temperature of the cryostat chamber should be checked. If there is no means of altering the temperature of the block during cutting, then the chamber has to be at a suitable temperature for the tissue and the type of preparation to be cut. Unfixed material sections well between −15°C and −23°C. Fixed tissues will section best within the range −7 to −12°C, depending upon the nature of the tissue.

4.4.2. Knife

Inaccurate sharpening of the knife can lead to problems if too large a facet is produced. The adjustment of the anti-roll plate is difficult and if the facet is large and the block small, difficulty is encountered in picking the section up onto a slide or cover-slip. The finish required on the knife edge is less than for paraffin-wax sectioning; a fine edge is rapidly removed by a hard frozen block of tissue and stropping the knife is not required. As an alternative to laboratory sharpening, knives can be re-ground commercially.

4.4.3. Anti-roll plate

This piece of equipment, attached to the front of the microtome in the majority of cryostats, is intended to stop the natural tendency of frozen sections to curl upwards on sectioning. The device is usually made of glass or Perspex coated with PTFE. The anti-roll plate is aligned parallel to the knife edge and fractionally above it. Adjustment is available to raise or lower the plate against the knife, as well as the angle between knife and plate. Its micro-adjustment determines the success of sectioning. The following points should be considered in relation to the anti-roll plate:

Correct height to knife edge
Correct angle to knife
Top edge not damaged
At cabinet temperature

4.4.4. Cryostat sectioning technique

Sectioning in a cryostat is straightforward if the points discussed above are followed. It is necessary to consider the speed of cutting; soft tissue sections better at a slow rate while harder tissue cuts better faster. Sectioning fixed tissue is more difficult and practice is needed to master this aspect of cryotomy.

To produce good sections, the knife and anti-roll plate should be cleaned of tissue debris and condensation before sectioning by using an artist's long hog-bristle brush. The cut section rests on the knife face and is picked up on a slide or cover-slip. It is a matter of preference which, but in the case of enzyme histochemistry the cost of reagents usually dictates the use of cover-slips. Sections of fixed blocks have a tendency to float off the slide or cover-slip during incubation or staining. This is avoided by coating the slides or cover-slips in a gelatine–formaldehyde mixture (see p. 59) before picking up the sections.

4.4.5. Containers for handling liquefied gases

It is important that liquefied gases are stored and transported safely and correctly. Specialized Dewar containers which have a capacity of up to 50 litres are commercially available, and large-capacity Dewars are suitable for storage of liquid nitrogen. *It has been reported that some workers use domestic vacuum flasks. This is an extremely dangerous practice and should be strongly discouraged.* Liquid nitrogen can leak between the inner glass flask and outer container, producing an increase in pressure when the liquid nitrogen evaporates which will cause the vacuum flask to explode. The use of a Dewar flask is essential when freezing tissue with isopentane super-cooled by liquid nitrogen.

4.4.6. Storage of tissues

Quenched tissues can be stored in the cryostat for a short period, but unless the tissue is kept air-tight by covering with Parafilm or cellophane, it will become dry and unusable. Prolonged storage should be either at $-70\,^{\circ}$C in a deep-freeze or in liquid-nitrogen specimen-storage containers. If the frozen tissue has been mounted onto a piece of cork during freezing, its subsequent handling will be easier. The specimen and cork are removed from the block-holder and stored in sealed polythene bags. The specimen can also be wrapped in aluminium foil or a piece of ice placed into the bag, to prevent the tissue from drying.

4.5. Ultracryotomy

During the last 20 years there have been many attempts to cut ultrathin (50–100 nm) frozen sections from unfixed unembedded tissue for electron microscopy. Ultracryotomy is a highly-specialized procedure and although many improvements have been made, there are many problems to be overcome. The complex and research nature of the topic in terms of equipment and techniques, especially enzyme histochemistry, make it a subject beyond the scope of this book. Readers interested should consult Christensen (1971) and Pearse (1980) for a general account; and Sjöstrand and Bernhard (1976), and Hayat (1977) for specialized procedures.

4.6. Freeze drying

Freeze drying is employed to avoid the disadvantages produced by fixation and processing. The drawbacks of these procedures may be summarized as follows:

Loss of soluble substances.
Displacement of cell constituents.
Chemical alteration of reactive groups.
Denaturation of proteins.
Destruction or inactivation of enzymes.

Freeze drying of biological tissues has been practised since its introduction by Altmann in 1890 but was not a practical proposition until Gersh in 1932 introduced his modifications. Pearse (1963) improved the technique with his *thermoelectric dryer* which is in use today. The freeze-drying technique involves rapidly quenching fresh tissue at −160°C, and subsequent removal of water molecules by sublimation in a vacuum at a higher temperature of −40°C. The tissue blocks are then raised to room temperature and fixed or embedded in a suitable medium. The purpose of the technque is to preserve the chemical composition and cell structure as near as possible to life. It is important that the dimensions of the tissue block are kept small (approximately 2 mm^3) if drying is to be satisfactory. Freeze drying can be considered in three stages: *Quenching; drying;* and *embedding and fixation*.

4.6.1. Quenching

This has three important effects. It stops the chemical reactions within the tissue; the importance of this to the histochemist is paramount. It brings the tissue to a solid state; in this condition diffusion of the tissue constituents is halted. The third effect is the formation of ice crystals by freezing of unbound water in the tissue. Damage is caused to the morphological appearances of the tissue if large crystals are formed, and the more rapid the freezing the smaller the ice crystals. In freeze drying this necessitates the use of freezing solutions of high thermal conductivity. Liquid nitrogen at −190°C has a low temperature but its conductivity is poor owing to the formation of vaporized gas around the tissue, slowing down the transfer of heat from tissue to freezing solution. Freon-22 has a high rate of thermal conductivity and if cooled to −160°C by placing in liquid nitrogen, suitably rapid freezing will be obtained. Other suitable quenching baths are Arcton-12 and iso-pentane.

4.6.2. Drying

This is the most time-consuming part of the method; some tissues contain 70–80 per cent water by weight and this has to be removed without damage to the tissue. Drying can be divided into three distinct stages (Meryman 1960).

Introduction of heat to the tissue to cause sublimation of ice.
The transfer of water vapour from ice crystals through the dry part of the tissue.
The removal of water vapour from the surface of the specimen.

Drying of tissue occurs when heat is supplied to the frozen tissue in a vacuum of 133 mPa or better. The heat vaporizes the water molecules and they pass through the tissue to the surface. There must be efficient removal of water molecules from the surface of the specimen. This is achieved by a vapour concentration gradient, having a high vapour pressure at the drying boundary and a low vapour pressure at the surface (Pearse, 1980).

Water molecules leaving the tissue are removed by a vapour trap. There are two types: a *cold finger trap* and a *chemical trap*. Cold finger traps were popular in the early type of freeze dryer. The chemical trap is used with the thermoelectric freeze dryer, with phosphorus pentoxide as the dehydrant.

4.6.3. Embedding and fixation

When the tissue is dry, it is brought to room temperature. At atmospheric pressure it will not absorb moisture unless drying is incomplete. Dried tissue is extremely friable and any undue pressure will cause the tissue to disintegrate. The delicate tissue may be fixed, but if enzyme histochemistry is required, post-fixation is often applied to the sections.

Following fixation or straight from the freeze dryer unfixed, the tissue is embedded in wax. A number of waxes can be used, paraffin usually gives acceptable results. Celloidin, followed by paraffin wax, is recommended for enzyme work (Burstone 1962). Polyester wax also gives good results in enzyme histochemistry. The practical aspects of using thermoelectric freeze dryers are dealt with in detail in Bancroft (1975) and Pearse (1980).

4.6.4. Demonstration of enzymes

The preservation of hydrolytic enzymes is good in freeze-dried tissue. Care has to be taken with the embedding medium; polyester wax with a melting point of $45\,°C$ is acceptable. If unfixed sections are used, considerable diffusion artefact may occur after incubation with poor results. Brief fixation in alcohol or acetone (1–2 h) is sufficient to reduce the subsequent diffusion of many enzymes. Dehydrogenases are not demonstrated in freeze-dried sections with or without post-fixation (Pearse, 1980). The double-embedding technique of Burstone (1962) gives excellent results for hydrolytic enzymes. A slightly modified version of this is as follows:

METHOD FOR DOUBLE EMBEDDING OF FREEZE DRIED TISSUE
(Burstone 1962)

1. Place freeze-dried tissue in pre-cooled acetone at $4\,°C$ for 4 h
2. Transfer tissue to pre-cooled absolute alcohol at $4\,°C$ overnight
3. Place tissue in pre-cooled alcohol–ether solution (1:1) at $4\,°C$ for 3 h
4. Transfer tissue to 5 per cent celloidin in alcohol–ether solution (1:1) at $4\,°C$ overnight.
5. Remove excess celloidin and transfer to chloroform at $22\,°C$ for 1 h

6. Transfer to fresh chloroform at 22°C for 1 h
7. Place tissue in molten paraffin wax at 56°C for 15 min
8. Transfer tissue to fresh molten paraffin wax at 56°C for 45 min
9. Embed and section in the usual way
10. Apply the histochemical method

4.7. Freeze substitution

This technique, first described by Simpson (1941), involves the rapid freezing of small pieces of tissue in a similar manner as freeze drying, and the substitution of the ice formed in the tissue by a dehydrant at sub-zero temperatures. Simpson suggested this technique to avoid the necessity of the expensive equipment used in freeze drying. There are two basic types of freeze substitution: ice of the quenched tissue is substituted in dehydrating fluid, or alternatively the tissue is dehydrated in a fluid containing a fixative. A general freeze-substitution technique for tissue was described by Feder and Sidman (1958). For enzyme localization and demonstration, *section freeze substitution* gives better results.

4.7.1. Section freeze substitution

This technique requires rapid freezing of the tissue to −160°C as described for freeze drying. Sections are cut at 8–10 μm in a cryostat and transferred to water-free acetone, cooled to −70°C for 12 h. The sections are floated onto cover-slips or slides and allowed to dry. For oxidative enzymes and 5-nucleotidase, it is essential the section is coated with cold celloidin (1 per cent in 30 parts ether, 30 parts ethanol and 40 parts acetone) chilled to −70°C (Pearse, 1980). After drying, the enzyme method is applied. Chemical analysis by Chang and Hori (1962) showed high levels of enzymes present including succinate dehydrogenase. However, the final results are often disappointing with considerable diffusion seen.

4.8. Paraffin wax sections

As described earlier, enzymes are labile and easily destroyed. Many factors are involved including the effects of fixation and heat and it is essential that these are controlled for the demonstration of enzymes. In producing paraffin-wax sections, tissue is fixed and a relatively high temperature is required during infiltration and embedding. There are also the effects of processing reagents. Such a combination of adverse factors results in a decrease or total loss of enzyme activity, including the majority of the more resistant hydrolytic enzymes. One of the most resistant enzymes is naphthol AS-D chloroacetate esterase and the method of Moloney *et al.* (1960) works well. Others which may be demonstrated are naphthyl acetate esterase and peroxidase, but other enzymes are not preserved using the standard processing schedule.

4.8.1. Modified processing schedule

One of the earliest modified processing methods using paraffin wax was by Gomori (1952) who used celloidin in a double-embedding procedure. The method is seldom used today. Varying degrees of success have been achieved when standard processing methods are altered (Burstone 1962; Schlake *et al.* 1978; Chilosi *et al.* 1981; and Fujimori *et al.* 1981). Dehydrating and clearing reagents should be as inert as possible to minimize enzyme loss, and used at 4°C. A low melting-point wax is used, as above 40°C considerable loss of enzyme activity will occur. Lower melting-point waxes will improve enzyme retention, but are difficult to section. The solutions, including wax, are used for comparatively short times compared with standard schedules and small pieces of tissue are required to enable well-processed blocks to be produced. The accuracy of localization of the enzyme is poor. The effect of different processing methods, including paraffin wax, on histochemical reactions was discussed by Bancroft (1966).

4.8.2. Enzyme activity in paraffin blocks

Enzyme loss occurs within paraffin blocks during storage, and consequently retrospective studies are disappointing. This loss is slowed if blocks are kept cold. The use of paraffin-wax sections has an extremely limited use in enzyme histochemistry. The technical difficulties and inconsistent results are such that, except for few occasions, paraffin-wax sections should not be seriously considered.

PREPARATION OF TISSUE FOR PARAFFIN WAX (Fugimori *et al.* 1981)
1. Fix in formalin acetone buffer (see p. 59) at 0-4°C for 2 h
2. Wash in gum sucrose at 0-4°C for 12 h
3. Wash in 0.1M cacodylate buffer + 0.03 per cent Triton X-100 at 0-4°C for 90 min
4. Dehydrate with 50 per cent acetone + 0.03 per cent Triton X-100 at 0-4°C for 30 min
5. 100 per cent acetone + 0.03 per cent Triton X-100 at 0-4°C for 30 min
6. 100 per cent acetone + 0.03 per cent Triton X-100 at 20°C for 30 min
7. Methyl benzoate at 20°C for 30 min
8. Paraffin wax at 60°C for 50 min
9. Embed
10. Cut 8-μm sections
11. Apply the histochemical method

4.9. Resin sections

There are three groups of resins classified according to their chemical composition and characteristics. They are acrylic, polyester, and epoxy resins of which there are sub-types within each group. In the early 1950s acrylic resins were used for enbedding in electron microscopy, but were later replaced by polyester and epoxy resins which proved superior. Today, epoxy resins are the most popular for electron microscopy; many of them are hydrophobic but there are water-soluble variants, e.g. Aquon, (Gibbons 1959) and Durcupan (Stäubli 1960). Aquon is not available commercially and has to be extracted from Epon-812 in the laboratory. Glycol methacrylate (GMA), which is a water-soluble acrylic resin, was also used for electron microscopy (Rosenberg *et al.* 1960). It was thought these water-soluble resins would be more suitable for histochemical studies because of their hydrophilic nature, but this has not been the case.

Using hydrophobic epoxy resins only simple basic dyes such as Toluidine Blue penetrate the resin, severely restricting the choice of light-microscopy staining techniques. Other methods may be employed if osmium tetroxide fixed sections are initially oxidized with hydrogen peroxide and the resin etched prior to staining with a solution such as sodium methoxide. These techniques are troublesome and many workers do not use staining methods other than Toluidine Blue. In contrast, Ashley and Feder (1966) noted several techniques applicable to paraffin sections that could also be applied to GMA-embedded tissue, without etching the resin. The acceptance that GMA allows good staining with routine techniques has gained popularity owing to its simplicity and in many cases thin resin sections stained by techniques other than Toluidine Blue are beneficial for diagnosis.

The development of GMA for thin resin sections for light microscopy was stimulated by Ruddell (1967) who introduced a resin-mix formula suitable for large blocks of tissue. GMA, which chemically is 2-hydroxyethyl methacrylate (HEMA), has a low viscosity aiding penetration into large blocks unlike epoxy resins. Since the end of the 1970s GMA can also be purchased in commercial kits, which have made handling easier and production of good polymerized blocks more reliable. In addition GMA, which contains the impurity methacrylic acid, has been further refined, decreasing background staining present when basic dyes are used. Another acrylic resin, LR White (London Resin Co.), not based on GMA, was introduced in 1981. Using these resins processing, embedding, and polymerization can be performed at low temperatures allowing some enzymes to be preserved. Fixed tissue is used but LR Gold has been specifically developed for unfixed tissue.

4.9.1. Demonstration of enzymes for light microscopy

As discussed earlier, enzymes are labile substances, some more than others. In order to demonstrate enzymes in resin sections an appreciation of the factors which influence their preservation and demonstration is required. Several studies have been published, whose fixation and processing schedules vary, but the general approach is similar. These include Higuchi *et al.* (1979), Beckstead and Bainton (1980), Horton *et al.* (1980), Burnett (1982), and Beckstead (1983).

4.9.1.1. Fixation

It is necessary to know which enzymes withstand fixation. Many oxidative enzymes are destroyed or depleted in fixed tissue, but several hydrolytic enzymes survive aldehyde fixation under controlled conditions. Paraformaldehyde is recommended by many workers in preference to glutaraldehyde or acrolein, although Beckstead and Bainton (1980) recommended a mixture of the three. Dawson (1972) suggested formol calcium for frozen sections, which is also suitable for tissues embedded in resin for enzyme demonstration. Some hydrolytic enzymes withstand fixation better than others, e.g. alkaline phosphatase is more resistant than acid phosphatase. Treatment of the tissue with a sucrose solution at $4°C$ after fixation, enhances enzyme demonstration.

4.9.1.2. Processing and polymerization of resin

Using GMA, tissues can be processed via the monomer which is miscible with water, but those of the LR series are processed via alcohol. The processing time depends on the size of tissue block. In some instances, overnight infiltration is necessary to achieve good processing, although prolonged dehydration in alcohol should be avoided as the activity of enzymes is decreased. Polymerization of acrylic resins is an exothermic reaction which if not controlled leads to enzyme loss in addition to tissue and block damage. To slow the rate of polymerization and minimize the heat produced, it is necessary to change the environmental conditions and the standard resin mix. Moulds employed for casting the blocks, are partially immersed in cold water at $4°C$. The amount of catalyst is decreased, and the ratio of resin to accelerator (or hardener) increased. The amount of both catalyst and accelerator must however be sufficient to produce a block hard enough to cut, otherwise it is possible that the resin will only gel and not solidify.

One of the characteristics of acrylic resins is that oxygen interferes with polymerization, and to produce good blocks this has to be excluded. This is easy if BEEM capsules are used by topping up with resin and covering with their lids. Using open plastic trays, air is removed either by vacuum or replaced by pumping in another gas, e.g. nitrogen into an enclosed chamber such as a desiccator.

4.9.1.3. Sectioning

Sectioning of blocks is carried out as described by the resin manufacturers, making sure the sections are floated out on to an ambient-temperature water-bath and allow to air dry. In enzyme histochemistry, the use of cover-slips is more suitable than slides and sections should be dried at room temperature. The use of an adhesive is recommended – 0.1 per cent poly-L-lysine has proved excellent in our hands.

4.9.1.4. Demonstration

Several hydrolytic enzymes are demonstrated using routine enzyme histochemical staining procedures, although in some cases the incubation times have to be extended. The commercial resin kits vary in their chemical composition and constituents and some enzymes are better demonstrated in specific resins. The greatly improved resolution and histological detail of resin sections compared with frozen sections, makes enzyme histochemistry highly satisfactory in terms of localization.

4.9.1.5. Unfixed tissue

A recent development in the field of acrylic resins has been the introduction of LR Gold. This has the property of allowing processing of unfixed tissue at $-25\,^{\circ}$C. Cold processing of unfixed tissue includes polyvinyl pyrollidone in the processing solutions to protect the tissue from osmotic changes. The resin is polymerized by light reacting with a photocatalyst mixed in the resin. The blocks are cast in gelatine capsules. Further appraisal of this resin is necessary before it can be unequivocally recommended. Thompson and Germaine (1984) have demonstrated the sensitive oxidative enzyme, succinate dehydrogenase, in addition to several hydrolytic enzymes.

PREPARATION OF TISSUE FOR GLYCOL METHACRYLATE (Beckstead 1983)
1. Fix tissue at $4\,^{\circ}$C in 4 per cent paraformaldehyde in 0.1 M phosphate buffer pH 7.4 for 2–6 h
2. Wash at $4\,^{\circ}$C in 0.1 M phosphate buffer pH 7.4 with 3 per cent sucrose for 1–12 h
3. Dehydrate in graded acetone at $4\,^{\circ}$C
4. Place tissue in acetone–glycol methacrylate (1:1) mixture at $4\,^{\circ}$C
5. Transfer to glycol methacrylate for overnight infiltration at $4\,^{\circ}$C
6. On following day, transfer tissue to embedding moulds containing the complete embedding mixture of 20 ml solution A (glycol methacrylate) + 100 mg benzoyl peroxide + 0.5 ml solution B (polyethylene-glycol-400 with N,N-dimethylaniline). These are components of a commercial kit, under the trade name of JB-4 (Polysciences Inc.)
7. Remove air by vacuum and allow resin to polymerize overnight at $4\,^{\circ}$C
8. Cut sections in the usual way and allow them to air dry at room temperature
9. Apply the histochemical method

4.9.2. Demonstration of enzymes for electron microscopy

Electron cytochemistry (*ultrahistochemistry*) is a rapidly expanding area whose origins go back to the early 1960s when the metal-precipitation principle formed the basis of the pioneer work. The current techniques, some of which are now more consistent and reliable, are not routine and are only performed in specialized research laboratories. The application of ultracryotomy for enzyme histochemistry has been previously discussed (p. 12) and although theoretically it remains the ideal technique, the technical difficulties have meant other procedures are used.

4.9.2.1. Fixation

Formol calcium is an acceptable fixative for enzyme histochemistry at the light-microscopic level. Although the use of formaldehyde is satisfactory for electron cytochemistry, Holt and Hicks (1961) reported that the inclusion of calcium was undesirable owing to the effect on fine structure. The introduction of glutaraldehyde by Sabatini *et al.* (1963) showed good preservation of ultrastructure. Several studies have been carried out on the effects of different fixatives on enzymes, revealing that the activity of each enzyme depends on the fixative employed. In general, glutaraldehyde or paraformaldehyde are chosen for electron cytochemistry, and whilst other fixatives allow greater enzyme activity, the electron histochemist has to compromise between preservation of enzyme activity and preservation of morphology. Osmium tetroxide used in electron microscopy as both a fixative and a 'stain' can destroy enzyme activity.

4.9.2.2. Visibility of enzymes

If an enzyme acts upon a specific substrate, and produces an electron-dense reaction product, or one which can subsequently be converted into an electron-dense material, then that enzyme can be demonstrated in the electron microscope. The metal-precipitation techniques of Gomori for acid and alkaline phosphatase both result in insoluble and highly electron-dense reaction products. Unfortunately, diffusion of the primary reaction product occurs and lead is often absorbed non-specifically on free tissue surfaces giving false localization. An alternative is the use of diazonium and tetrazolium salts linked to osmium. An example of this type of technique was described by Tsou *et al.* (1968) who utilized the osmate of 2,2′,5,5′-tetra-*p*-nitrophenyl-3,3′-stilbene ditetrazolium chloride (Os-TNST) to demonstrate succinate dehydrogenase activity in cristae of mitochondria. In addition to naturally-occurring electron-dense products, Hanker *et al.* (1971) reported that if copper ferrocyanide (Hatchett's Brown) produced either by cholinesterase or succinate dehydrogenase is converted to the sulphide, and reacted with osmium, an osmiophilic polymer can form which greatly enhances electron density and contrasts. This principle is described as *osmiophilic polymer generation*. Diaminobenzidine (DAB) is osmiophilic and this has been utilized in the demonstration of phosphatases, esterases, peroxidases, and dehydrogenases. Osmication is achieved either by vapour or a solution of osmium tetroxide.

4.9.2.3. Tissue preparation for ultrahistochemistry

The application of enzyme techniques on resin-embedded sections is impractical, because enzymes are unable to withstand the effects of processing and polymerization of epoxy resins. To overcome these problems, the enzyme histochemical technique is applied before the tissue section or block is embedded in resin. There are three main ways to produce material suitable for subsequent histochemistry: *Frozen sections; small tissue blocks; non-frozen sections.*

The block or section can be fixed or unfixed before the histochemical method is applied, but to improve enzyme localization and aid handling, fixation is usual.

Frozen sections

Tissue is frozen rapidly and sections cut either on a freezing microtome or in a cryostat (Boadle and Bloom 1969) at approximately 50 μm. The histochemical method is performed and tissue processed into resin for subsequent ultrathin sectioning. This procedure has been applied to both fixed and unfixed tissue, but the latter suffer from the problems of difficult handling and poor localization of the enzyme. Unfortunately, freezing of fixed tissue causes greater structural damage than freezing of unfixed tissue.

Small tissue blocks

Small blocks approximately 1 mm^3 are incubated in the histochemical medium before being processed to resin for sectioning (Slezak and Geller 1984). The problem with this technique is the poor penetration of the incubating medium into a comparatively thick piece of tissue.

Non-frozen sections

The use of non-frozen sections produced by tissue-chopping or slicing instruments such as a vibratome is the most common approach for ultrahistochemistry. Non-frozen sections of up to 75 μm are incubated in a histochemical medium before processing to resin for ultramicrotomy. This method does not produce architectural artefacts associated with freezing and the penetration of the incubating medium is satisfactory. Tissue sections are essential to avoid non-specific staining and non-frozen sections produce sharper ultrastructural localization of reaction products.

4.9.2.4. Enzyme-demonstration techniques

There are an increasing number of techniques for the demonstration of enzymes at the ultrastructural level. Early methods were based on existing light-microscopy techniques using the metal-precipitation techniques, but later modifications especially with diazonium and tetrazonium salts were tried. Recently, new methods have been introduced including those which utilize cerium as a capture agent. This method was first applied to the fine structural demonstration of NADH oxidase by Briggs *et al.* (1975) and has been used in the localization of many enzymes including monoamine oxidase (Fujimoto *et al.* 1982) and several phosphatases (Robinson and Karnovsky 1983, and Hulstaert *et al.* 1983). There are numerous techniques available for enzyme ultrahistochemistry. Further details are given by Shnitka and Seligman (1971), Pearse (1972), Hayat (1973, 1974a, 1974b, 1975, 1977), and Lewis and Knight (1977).

PREPARATION OF TISSUE FOR THE ULTRACYTOCHEMICAL DEMONSTRATION OF PHOSPHATASES BY A CERIUM-BASED METHOD (Hulstaert *et al.* 1983)

1. Perfuse the tissue *in situ* with 2 per cent polyvinylpyrrolidone, 75 mM sodium nitrite in 0.1 M cacodylate buffer pH 7.4 for 1 min
2. Perfuse with 2 per cent glutaraldehyde in 0.1 M cacodylate buffer pH 7.4 for 5 min
3. Perfuse with 6.8 per cent saccharose in 0.1 M cacodylate to remove fixative
4. Cut 30-μm sections on a vibratome
5. Incubate sections in appropriate enzyme histochemical medium
6. Rinse sections at 4 °C in 6.8 per cent saccharose in 0.1 M cacodylate buffer pH 6.0 overnight
7. Post-fix sections at 4 °C in 1 per cent osmium tetroxide, 1.5 per cent potassium ferrocyanide in 0.1 M cacodylate buffer pH 7.4 for 2 h
8. Rinse sections in 6.8 per cent saccharose in 0.1 M cacodylate buffer pH 7.4
9. Dehydrate sections in graded alcohol
10. Embed in Epon-812
11. Cut and view ultrathin sections in the usual manner

4.10. Smears

During the development of enzyme histochemistry, there have been several reports of the use of cytological smears for cytochemical identification and evaluation of cells. Such smears, whose preparation may be achieved in various ways, include: blood (Wachstein and Wolf 1958, Golberg and Barka 1962, Li *et al.* 1970, Yam *et al.* 1971, Catovsky *et al.* 1978); bone marrow (Wachstein and Wolf 1958, Yam *et al.* 1971, Catovsky *et al.* 1978); and tissue cell suspensions (Meuller *et al.* 1975).

Yam *et al.* (1971) describe several enzyme techniques for cytochemical identification of monocytes and granulocytes. Three of the most useful enzymes are non-specific esterase, acid phosphatase, and chloroacetate esterase. According to Higgy *et al.* (1977), non-specific esterase activity at pH 8.0 is seen in mature T-lymphocytes as a dense focal dot in the cytoplasm, whilst B-lymphocytes are negative, and monocytes show a diffuse pattern of granular positivity. At pH 6.3 monocytes become negative (Reynolds 1982). Acid phosphatase activity is found in many cells including granulocytes, monocytes, and T-lymphocytes, although their distribution and staining pattern differ. Higgy *et al.* (1977) reported that acid phosphatase failed to differentiate between T- and B-lymphocytes, but Wright and Isaacson (1983) reported weaker staining in B-cells. The intensity may be decreased in neoplastic cells compared with normal cells. Acid phosphatase present in T-cell lymphomas is tartrate-labile, whereas in hairy-cell leukaemia the enzyme is tartrate-resistant. Lojda (1977 and 1981) describes the selective staining of T-lymphocytes on blood smears by the demonstration of the proteinase, dipeptidyl(amino)peptidase-IV (DAP-IV). Chloroacetate esterase is used as a marker for granulocytes including promyelocytes and many myeloblasts. A summary of the histochemical reactions of myeloid and lymphoid cells is shown in Table 2.

Table 2. *Histochemical reactions of myeloid and lymphoid cells.*

Cell type	AP	NSE	CAE	DAP–IV
Granulocytes	D	–	D	–
Monocytes/macrophages	D	D	±	–
Immature T-cells	F	–	–	D
Mature T-cells	F	F	–	D*
B-cells	±	–	–	–

Modified from Wright and Isaacson (1983).
* T-helper cells; D: diffuse staining; F: focal staining; AP: acid phosphatase; NSE: non-specific esterase; CAE: chloroacetate esterase; DAP–IV: dipeptidyl(amino)peptidase–IV.

4.10.1. Imprint smears

The technique of preparing imprint smears is simple and rapid, and provides valuable diagnostic information if carried out correctly. Fresh tissue is cut and the new surface touched gently against a clean glass slide avoiding excess pressure as this distorts cells. The slides are rapidly air-dried and stored unfixed at −70°C if staining is not immediately required. Lymph nodes may be handled in this manner.

4.10.2. Fixation

It is usual to fix smears before histochemical staining, to preserve cell structure and enzyme localization. Various fixatives have been used, with a formalin variant being the most popular and successful. Higgy *et al.* (1977) found formalin vapour superior to either paraformaldehyde or glutaraldehyde solutions for the demonstration of non-specific esterase and acid phosphatase, but many histochemists find cold (4°C) formol calcium satisfactory.

PREPARATION OF BLOOD SMEARS FOR ESTERASE AND ACID PHOSPHATASE CYTOCHEMISTRY (Higgy *et al.* 1977)
1. Prepare buffy coat smear from heparinized blood
2. Allow to air dry
3. Expose smear to formalin vapour for 4 min
4. Wash briefly in distilled water and blot dry
5. Apply the histochemical method

5

Demonstration of enzymes

5.1. Introduction

The basic principle of enzyme histochemistry is for the enzyme in the tissue to be presented with a solution containing a specific substrate. If, as a result of the catalytic action of the enzyme on this substrate, a coloured insoluble reaction product (RP) is produced, then the demonstration of the enzyme will have been achieved. Often however, the primary reaction product (PRP) is colourless, and requires coupling with a visualizing agent to generate a coloured insoluble final reaction product (FRP). Most enzymes rely on three basic techniques for their histochemical demonstration: *Metal precipitation*; *diazonium salts*; *tetrazolium salts*.

5.2. Techniques

5.2.1. Metal precipitation

This technique is applied to the demonstration of phosphatases. Phosphate ions, released as a result of enzyme activity on the substrate, combine with a suitable metallic cation to produce an insoluble precipitate of metal phosphate. The phosphates produced are invisible and are rendered visible by converting to a black sulphide. Metallic cations used for combining with released phosphate are calcium and lead.

5.2.2. Diazonium salts

Diazonium salts are prepared by treating primary aromatic amines with an acid solution of sodium nitrite. The resulting salt, generally a chloride, will react with enzymatically-released naphthol from the substrate to produce an intensely-coloured insoluble azo dye. The basic structure of diazonium salts is $R-N^+\equiv N$ which on formation to azo groups $(-N=N-)$ confer colour.

Diazonium salts do not keep indefinitely, and should be stored in a cool dark place and renewed at six-monthly intervals. During incubation, it is important to choose conditions carefully so the combination of reactants takes place as rapidly as possible, and the product formed remains insoluble through subsequent processing. Failure in either of these respects reduces the accuracy of localization of enzyme activity.

The rate of coupling of diazonium salts depends upon their chemical nature and the pH of the incubating medium. These factors determine to a large extent the choice of diazonium salt. Precision in locating the site of enzyme activity also depends upon the concentration of diazonium salt. There is an optimal concentration in the incubating solution; if the concentration is less, diffusion of the PRP may occur. If the concentration is greater than optimal, then inhibition of enzyme activity occurs, and the risk of non-specific background staining increases.

5.2.2.1. Simultaneous coupling using diazonium salts

This reaction utilizes an incubation solution containing the substrate and a diazonium salt in a suitable buffer. The enzyme present in the section hydrolyzes the substrate to form an invisible PRP which immediately couples with a diazonium salt to produce the coloured FRP. The substrate has to be soluble in water or in buffer to allow sufficient substrate to be available in solution for the enzyme to hydrolyze. The method is performed at the pH at which the enzymes exhibit maximum activity.

5.2.2.2. Post-coupling using diazonium salts

In this type of reaction the enzyme hydrolyzes the substrate producing a reasonably insoluble colourless PRP. Subsequent coupling with a diazonium salt to produce a coloured FRP is carried out in a separate solution. This type of procedure relies on the PRP remaining at the initial site of hydrolysis without diffusion. The absence of the diazonium salt with the substrate has several theoretical advantages. The optimum pH for the initial enzyme substrate reaction is attained in the first incubating solution, followed by a different optimum pH for coupling the PRP with the diazonium salt. The method also avoids any deleterious effects that diazonium salts may have on the enzyme–substrate reaction. In addition, long exposure of tissue to diazonium salts may produce non-specific staining. In practice, however, enzyme localization is poor unless the PRP is extremely insoluble avoiding any diffusion. The technique is seldom used, as suitable substrates are not available and the method has the risk of giving false-negative results. Coupling with diazonium salts can be represented thus:

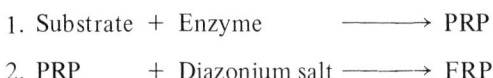

1. Substrate + Enzyme ⟶ PRP

2. PRP + Diazonium salt ⟶ FRP

In simultaneous coupling the substrate and diazonium salt are present in the same incubating medium; in post-coupling they are present in separate solutions.

5.2.3. Tetrazolium salts

The demonstration of many oxidative enzymes uses tetrazolium salts. These salts, which are colourless and water-soluble, accept hydrogen released from the substrate by enzyme action, and on reduction form highly-coloured water-insoluble microcrystalline deposits known as formazans.

Tetrazolium salts can form reversible combinations permitting electron transport

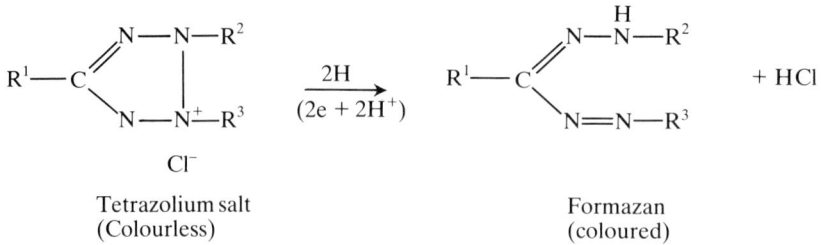

Tetrazolium salt	Formazan
(Colourless)	(coloured)

in situ to produce a formazan deposit at the site of enzyme activity. Several factors influence the characteristics and the selection of tetrazolium salts including light sensitivity, formazan crystal size, lipid solubility, and speed of reduction. Two types of tetrazolium salts have been developed: *monotetrazolium salts* and *ditetrazolium salts*.

The monotetrazolium salt, 3-(4,5-dimethyl-thiazolyl-2)-2,5-diphenyl tetrazolium bromide or methyl-thiazolyldiphenyl tetrazolium (MTT) was introduced by Pearse in 1957. The formazan produced chelates with cobalt ions present in the incubating medium. The final deposit is deeply coloured, finely granular, and is soluble in lipid. Using this salt, Scarpelli *et al.* (1958), Hess *et al.* (1958), and Pearse (1972) described methods for several dehydrogenases.

A ditetrazolium salt introduced by Tsou *et al.* (1956), 2,2'-di-p-nitro-phenyl-5,5'-diphenyl-3,3'-(3,3'-dimethoxy-4,4'-biphenylene) ditetrazolium chloride, is known as nitro blue tetrazolium (NBT). It is used for the demonstration of succinate dehydrogenase and forms a highly-coloured microcrystalline formazan which is insoluble in lipid.

The measurement of the ease, and consequently the suitability, of a tetrazolium salt to be reduced is known as the *redox potential*. It is measured in volts using a potentiometer.

5.3. Factors affecting demonstration

5.3.1. Temperature

The optimal temperature for the majority of enzyme reactions is $37\,^{\circ}\text{C}$. At higher temperatures the enzymes, being proteins, are typically denatured rapidly. Lower temperatures between 4 and $30\,^{\circ}\text{C}$ may be useful in demonstrating active enzymes as better localization is achieved by decreasing the rate of the enzyme reaction.

5.3.2. pH

The majority of enzymes have a pH at which the rate of the reaction is optimal. For many enzymes, this is around pH 7.0. Alkaline phosphatase at pH 9.2 and acid phosphatase at pH 5.0 are examples of exceptions to this rule.

5.3.3. Inhibitors

Enzyme activity can be destroyed by using chemical substances known as *inhibitors* which are classified as either *specific* or *non-specific*.

Specific inhibitors compete for essential sites on the enzyme molecule, whereas non-specific inhibitors prevent the reaction by denaturing the protein enzyme. Specific inhibitors are further sub-divided into *competitive* and *non-competitive*, depending on the location of the site of attack on the enzyme. Competitive inhibitors compete for the active site, i.e. the same site as the substrate, whereas non-competitive inhibitors affect the enzyme by binding to sites elsewhere. An example of a non-specific inhibitor is heat (which inhibits all enzymes); phosphate ions inhibit phosphatases competitively, and cyanide is an example of a non-competitive inhibitor.

5.3.4. Cofactors

To promote enzyme activity, the presence of chemicals referred to as *cofactors* is frequently required. These are often metal ions, e.g. magnesium ions are used in the Gomori metal-precipitation technique for alkaline phosphatase. In this role, they are called *activators*. Dehydrogenases are enzymes which oxidize their substrate by removing hydrogen and passing it to a suitable acceptor. The presence of coenzymes such as nicotinamide adenine dinucleotide (NAD) or nicotinamide adenine dinucleotide phosphate (NADP) in the incubating medium acts as the hydrogen acceptor. Reduced NAD and NADP are signified as NADH and NADPH, respectively. Coenzymes cannot be considered to be activators since they take an active part in the reaction catalysed.

5.4. The use of controls

Controls are a necessary part of enzyme histochemistry. Substrates, diazonium, and tetrazolium salts and other chemical solutions deteriorate with time, leading to possible failure of the method or occasionally false-positives. A positive control is carried though when incubating test sections.

Omission of the substrate from the incubating medium or the inclusion of specific inhibitors will act as a negative control. The absence of a reaction product indicates that positive results obtained with the authentic technique are yielding useful information about the distribution of the enzyme. Non-specific background staining is also excluded as a possible source of error if control sections give negative results.

6

Hydrolytic enzymes

Hydrolytic enzymes are numerous and have the ability to hydrolyze substrates by either the addition or removal of water. For ease these enzymes will be considered in the following groups: *Phosphatases*; *esterases*; *miscellaneous*.

Preparation of tissues for the demonstration of hydrolytic enzymes has previously been discussed in Chapter 3. Fixation in formol calcium followed by washing in sucrose is usually satisfactory. Unfixed frozen sections may also be used, although better enzyme localization is achieved if the tissue is fixed. Some enzymes, e.g. myofibrillar adenosine triphosphatase, require fresh unfixed frozen sections. Methods given which do not specifically require unfixed tissue should be assumed to also be applicable to fixed tissue.

6.1. Phosphatases

Phosphatases are present in a wide variety of animal tissues and are responsible for the hydrolysis of organic phosphate esters. Some act specifically on a substrate and are known as *specific phosphatases*. An example is adenosine triphosphatase which specifically hydrolyzes adenosine triphosphate. The remainder, whose substrate specificity is less limited, are divided into two groups: those exhibiting optimal activity at high pH values (*alkaline phosphatses*) and those exhibiting optimal activity at low pH values (*acid phosphatases*).

The hydrolysis of organic phosphate esters during incubation provides the basis for histochemical reactions. The released phosphate ions or the remaining organic residue is made visible by a variety of means. Phosphate ions can be precipitated as insoluble salts, i.e. lead or calcium phosphate, and converted to coloured sulphides which are visible. This is the basis of the many Gomori-type metal-precipitation techniques. Alternatively, the alcoholic residue of the substrate after enzymatic hydrolysis, can react with a diazonium salt to produce a highly-coloured insoluble azo dye. Phosphate esters of α-naphthol or its derivatives are the most commonly-used substrates. Coupling may take place during incubation (simultaneous coupling) or after incubation (post-coupling). The techniques are often referred to as *azo-dye methods*.

For the demonstration of phosphatases, Pearse (1960) states that the average optimal concentration of the diazonium salt is 1 mg/ml of incubating solution. If the concentration is less, diffusion of the primary reaction product, α-naphthol, occurs. On the other hand, if the concentration is much greater, say 5 mg/ml of incubating medium, inhibition of primary reaction product formation occurs, and non-specific background staining is seen. Adequate controls should be used to eliminate the possibility of failure of the method, and of producing false-positives which can occur if calcium or iron is present in the tissues.

6.1.1. Alkaline phosphatases

These enzymes exhibit optimum activity at an alkaline pH, and are widely distributed. They are activated by magnesium, manganese, and cobalt ions. Cyanide and cysteine inhibit alkaline phosphatase activity and may be incorporated in the incubating medium to provide a control.

CALCIUM PHOSPHATE METAL-PRECIPITATION METHOD FOR ALKALINE PHOSPHATASE (Gomori 1952, modified)

Original methods for the demonstration of alkaline phosphatases were described independently by Gomori and Takamatsu in 1939. A variation of the original Gomori technique published in 1952 is given below.

If a section is placed in an incubating medium containing the substrate (sodium β-glycerophosphate), calcium ions (calcium nitrate), and an activator for the phosphatase (magnesium chloride), a precipatate of calcium phosphate is formed at the sites of enzyme activity. The enzyme liberates phosphate ions from sodium β-glycerophosphate which combine with calcium ions to form calcium phosphate. This precipitate is treated with cobalt nitrate to produce cobalt phosphate, which is reacted with dilute ammonium sulphide to form cobalt sulphide. This is visible as a black deposit. The reactions can be summarized as follows:

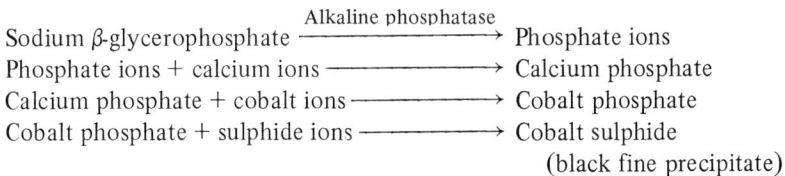

Sodium β-glycerophosphate $\xrightarrow{\text{Alkaline phosphatase}}$ Phosphate ions
Phosphate ions + calcium ions $\longrightarrow$ Calcium phosphate
Calcium phosphate + cobalt ions $\longrightarrow$ Cobalt phosphate
Cobalt phosphate + sulphide ions $\longrightarrow$ Cobalt sulphide
(black fine precipitate)

The incubating medium does not keep and it is advisable to make up fresh solutions immediately before use. The final pH of the solution is critical. If pH values less than 9.0 are employed, enzyme activity is impaired and the intensity of the final reaction product is reduced. Furthermore, localization is adversely affected by the solubility of calcium phosphate at a lower pH. The duration of incubation is determined by experiment and varies with the nature of the tissue. Long incubation times are to be avoided as this favours diffusion of the enzyme.

Preparation of incubating medium

2 per cent Sodium β-glycerophosphate	2.5 ml
2 per cent Sodium barbitone	2.5 ml
2 per cent Calcium nitrate	5.0 ml
1 per cent Magnesium chloride	0.25 ml
Distilled water	1.25 ml

The final pH of the incubating medium should be between 9.0 and 9.4.

Method

1. After suitable fixation, bring sections to water, incubate in the above medium at 37°C for 10 min to 6 h
2. Wash well in distilled water
3. Repeat wash
4. Treat section with 2 per cent cobalt nitrate for 3 min
5. Wash well in distilled water
6. Repeat wash
7. Immerse sections in 1 per cent ammonium sulphide for 2 min
8. Wash well in distilled water
9. Counterstain in 2 per cent methyl green (chloroform-washed)
10. Wash well in running tap water
11. Mount in glycerine jelly

Results

Alkaline phosphatase activity appears brownish-black and nuclei are green.

It is convenient to keep the stock solutions made up in batches. The incubation time varies according to the type of preparation; frozen sections requiring the shortest time.

SIMULTANEOUS COUPLING METHOD USING A DIAZONIUM SALT FOR ALKALINE PHOSPHATES (Gomori 1951)

This method was first described by Menton *et al.* (1944). The technique has been modified and improved upon by a number of people, notably by Gomori (1951). It employs sodium α-naphthyl phosphate as the substrate, together with a suitable diazonium salt buffered to pH 9.2. The enzyme liberates α-naphthol from the substrate, which is subsequently coupled to the diazonium salt producing an insoluble azo-dye at the site of enzyme activity.

Preparation of incubating medium

Sodium α-naphthyl phosphate	10 mg
0.2 M Tris buffer (stock solution A only), pH 10.0	10 ml
Diazonium salt (Fast Red TR)	10 mg

The final pH of incubating medium should be between 9.0 and 9.4. Sodium α-naphthyl phosphate is dissolved in buffer, the diazonium salt is added, and the solution well mixed. The solution is then filtered and used immediately.

Method

1. After fixation, bring sections to water, incubate in the above medium at room temperature for 10–60 min
2. Wash in distilled water
3. Counterstain in 2 per cent methyl green (chloroform-washed)

4. Wash in running tap water
5. Mount in glycerine jelly

Results

Alkaline phosphatase activity appears reddish-brown and nuclei are green.

SIMULTANEOUS COUPLING METHOD USING SUBSTITUTED NAPHTHOLS FOR ALKALINE PHOSPHATASE (after Burstone 1958a)

The use of substituted naphthols for the demonstration of alkaline phosphatases was first studied in detail by Burstone. The substituted naphthol esters he studied included AS-BI, AS-MX, and AS-TR phosphate. These esters are hydrolyzed rapidly by alkaline phosphatases yielding extremely insoluble naphthol derivatives which react with a diazonium salt to produce an insoluble azo-dye at the site of activity. The localization with substituted naphthols is superior to those obtained using alternative methods. Substituted naphthols also couple with diazonium salts over wider pH ranges.

Stock solutions of substituted naphthol phosphates may be produced by dissolving the compound in N,N-dimethylformamide (DMF) and adding distilled water and pH 8.3 buffer. These solutions keep in a refrigerator for several months. To apply the method add a suitable diazonium salt (e.g. Fast Red TR, Fast Blue RR) to the required amount of incubating solution in the ratio of 1 mg/ml. The solution should be prepared immediately before use.

Preparation of solutions

(a) Naphthol AS-BI stock solution

Naphthol AS-BI phosphate	25 mg
N,N-dimenthylformamide	10 ml
Distilled water	10 ml
Molar sodium carbonate	2–6 drops

The reagents are added in the above order, sufficient molar sodium carbonate is added until the pH is 8.0; then add

Distilled water	300 ml
0.2 M Tris-HCl buffer, pH 8.3	180 ml

The solution, which is faintly opalescent, is stable for many months.

(b) Incubating medium

Stock naphthol AS-BI solution	10 ml
Fast Red TR	10 mg

Shake well, filter, and use immediately.

Method

1. After fixation, bring sections to water, incubate in the above medium at room temperature for 5–15 min

2. Wash in water
3. Counterstain in 2 per cent methyl green (chloroform-washed).
4. Wash well in running tap water
5. Mount in glycerine jelly

Results

Alkaline phosphatase activity appears as red, and nuclei are green.

There are two points to watch: The addition of molar sodium carbonate changes the pH rapidly and care must be taken not to make the solution too alkaline. The reaction is fast, and care must be taken not to over-incubate.

6.1.2. Acid phosphatases

These enzymes received less attention originally than alkaline phosphatases owing to the fact that only in recent years have the methods for their demonstration become reliable. The problem is the solubility of acid phosphatases, and the difficulty of obtaining accurate localization of the final reaction product. Simultaneous coupling azo-dye methods are also beset by the difficulty of finding diazonium salts that couple efficiently under the acid conditions necessary for optimal enzyme activity. The enzymes are distributed widely throughout the body, the kidney, liver, and spleen being particularly rich. Fluoride ions inhibit these enzymes, and the inclusion of sodium fluoride in the incubating medium affords a reliable control measure.

LEAD PHOSPHATE METAL PRECIPITATION METHOD FOR ACID
PHOSPHATASE (Gomori 1941)

This method is similar to Gomori's method for alkaline phosphatases. Calcium phosphate which is formed in the alkaline phosphatase technique cannot be used owing to its solubility at acid pH levels. Lead nitrate is therefore used to precipitate the phosphate ions. The section is incubated in a medium containing sodium β-glycerophosphate (substrate) and lead nitrate; no activator is required. The enzyme splits phosphate ions from the substrate and these form an insoluble precipitate of lead phosphate at the site of enzyme activity. The sections are then treated with a dilute solution of ammonium sulphide to form lead sulphide, seen under the microscope as a dense brown deposit. The deposit is granular; if the optimal conditions for the method have been fulfilled, the granules should be small. The reaction can be summarized as follows:

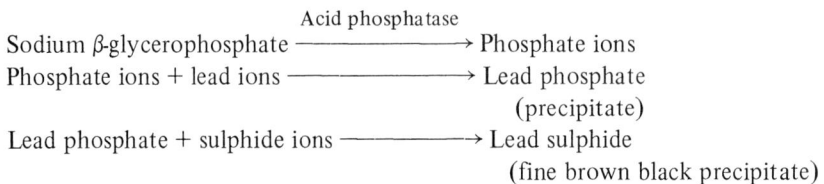

$$\text{Sodium } \beta\text{-glycerophosphate} \xrightarrow{\text{Acid phosphatase}} \text{Phosphate ions}$$

$$\text{Phosphate ions + lead ions} \longrightarrow \text{Lead phosphate}$$
$$\text{(precipitate)}$$

$$\text{Lead phosphate + sulphide ions} \longrightarrow \text{Lead sulphide}$$
$$\text{(fine brown black precipitate)}$$

The incubating medium is made up fresh for each batch of sections. It is convenient to keep the individual reagents in stock solutions stored at 4°C. It is important to use reagents of a high purity to obtain satisfactory results. The pH of the final incubating solution is important; the method works best at a pH of 4.8–5.0. If the pH is increased much above 5.5, it is possible to demonstrate other phosphatases.

Preparation of incubating medium

0.1 M Acetate buffer, pH 5.0	10 ml
Lead nitrate	20 mg
Sodium β-glycerophosphate	32 mg

The lead nitrate is dissolved in the buffer before the sodium β-glycerophosphate is added. The pH of the incubating medium should be approximately 5.0.

Method

1. Place sections in the above incubating medium at 37°C for 0.5–2 h
2. Wash in distilled water
3. Immerse for 2 min in 1 per cent ammonium sulphide (fresh)
4. Wash well in distilled water
5. Counterstain in either 2 per cent methyl green (chloroform-washed) or Mayer's carmalum
6. Wash in tap water
7. Mount in glycerine jelly

Results

Acid phosphatase activity appears black, and nuclei are green or red.

SIMULTANEOUS COUPLING METHOD USING A DIAZONIUM SALT FOR ACID PHOSPHATASE (after Grogg and Pearse 1952)

The principle employed in the corresponding technique for alkaline phosphatase is used for the demonstration of acid phosphatase. The substrate sodium α-naphthyl phosphate (1-naphthyl phosphoric acid) is dissolved in 0.1 M veronal acetate buffer, pH 5.0. Naphthol is released at the site of enzyme activity and coupled with a suitable diazonium salt. The problem with azo-dye methods for acid phosphatases is finding a diazonium salt to couple satisfactorily at an acid pH. A comprehensive study of different diazonium salts and their performance when used in this technique was described by Pearse (1972). Fast Garnet GBC gave the best results.

Preparation of incubating medium

Sodium α-naphthyl phosphate	10 mg
0.1 M Acetate buffer, pH 5.0	10 ml
Fast Garnet GBC	10 mg

Sodium α-naphthyl phosphate is dissolved in buffer and the diazonium salt added. The solution is filtered and used immediately.

Method

1. Incubate in the above medium at 37 °C for 15–60 min
2. Wash in distilled water
3. Counterstain in 2 per cent methyl green (chloroform-washed)
4. Wash in running tap water
5. Mount in glycerine jelly

Results

Acid phosphatase activity appears red, and nuclei are green.

SIMULTANEOUS COUPLING METHOD USING SUBSTITUTED NAPHTHOLS FOR ACID PHOSPHATASE (modified by Barka 1960)

The substituted naphthols Burstone studied also couple efficiently at an acid pH. He recommended naphthol AS-BI phosphate as the one of choice for acid phosphatase (Burstone 1958b) owing to the insolubility of the reaction product. Barka (1960) recommended the use of hexazonium pararosaniline as the diazonium salt, first used by Davis and Ornstein (1959). According to Barka, hexazonium pararosaniline does not inhibit the enzyme at an acid pH, and he attributed the improved localization obtained using this salt to the extreme insolubility and substantivity of the azo-dye produced. The combination of hexazonium pararosaniline and naphthol AS-BI phosphate allows accurate localization of the reaction product and is recommended.

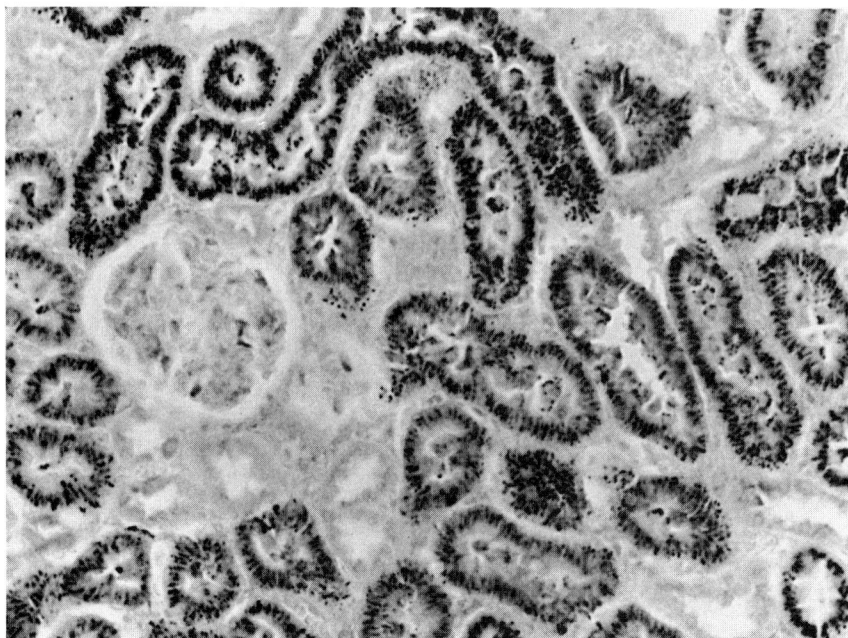

Fig. 1. Section of mouse kidney demonstrating acid phosphatase. The method utilizes a substituted naphthol (AS-BI) and pararosaniline – HCl as the diazonium salt. (×290)

The diazonium salt is prepared in two stages. Pararosaniline hydrochloride is dissolved in distilled water and acidified with concentrated hydrochloric acid. The solution is filtered and stored at room temperature and is stable for several months. Diazotization is achieved by the addition of 4 per cent sodium nitrite. It is important this solution is freshly prepared. Incubation is carried out at room temperature for periods of 30 min to 2 h, or at 37°C for 10–30 min. The sections may be dehydrated rapidly through graded alcohols to xylene, and mounted in DPX.

Preparation of solutions

(A) Substrate solution

Naphthol AS-BI phosphate	10 mg
N,N-dimethylformamide	1 ml

(B) Buffer solution

Sodium acetate $3H_2O$	1.17 g
Sodium barbitone	2.94 g
Distilled water	100 ml

(C) 4 per cent Sodium nitrite

(D) Pararosaniline HCl stock solution

Pararosaniline	1 g
Distilled water	20 ml
Hydrochloric acid (conc.)	5 ml

(E) Distilled water

Preparation of incubating medium

Solution (A)		0.5 ml
Solution (B)		2.5 ml
Solution (C)	⎰ Mix 0.4 ml of solutions	
Solution (D)	⎱ (C) and (D) first	0.8 ml
Solution (E)		6.5 ml

For the success of the technique, equal parts of solutions (C) and (D) are mixed together and allowed to stand for 2 min before being added to the incubating medium. The final pH should be between 4.7–5.0; it is adjusted if necessary with 0.1 M NaOH. For tartrate inhibition studies, 75 mg L-tartaric acid should be added to each 10 ml of incubating medium before adjusting the pH. Filter the solution before use.

Method

1. Incubate sections in the above medium at 37°C for 10–30 min
2. Wash in distilled water
3. Counterstain in 2 per cent methyl green (chloroform-washed)
4. Wash in running water

5. Either mount in glycerine jelly, or dehydrate rapidly through fresh alcohols to xylene and mount in DPX.

Results

Acid phosphatase activity appears red, and nuclei are green.

Note

Appropriate aliquots of solutions (C) and (D) may be stored frozen for convenience.

6.1.3. Adenosine triphosphatases

These are specific phosphatases which hydrolyze adenosine triphosphatase. There are several types, but the most important adenosine triphosphatases (ATPases) are: *Myofibrillar ATPase; membrane ATPase; mitochondrial ATPase.*

The demonstration of ATPases is determined by the technique selected as each is influenced by specific conditions. Only the method for myofibrillar ATPase modified by Brooke and Kaiser (1970) is described which reveals different fibre types in skeletal muscle. At pH 9.4, ATPase splits off phosphate from ATP which combines with calcium in the incubating medium to form calcium phosphate. Further reactions follow the rationale of the metal precipitation method for alkaline

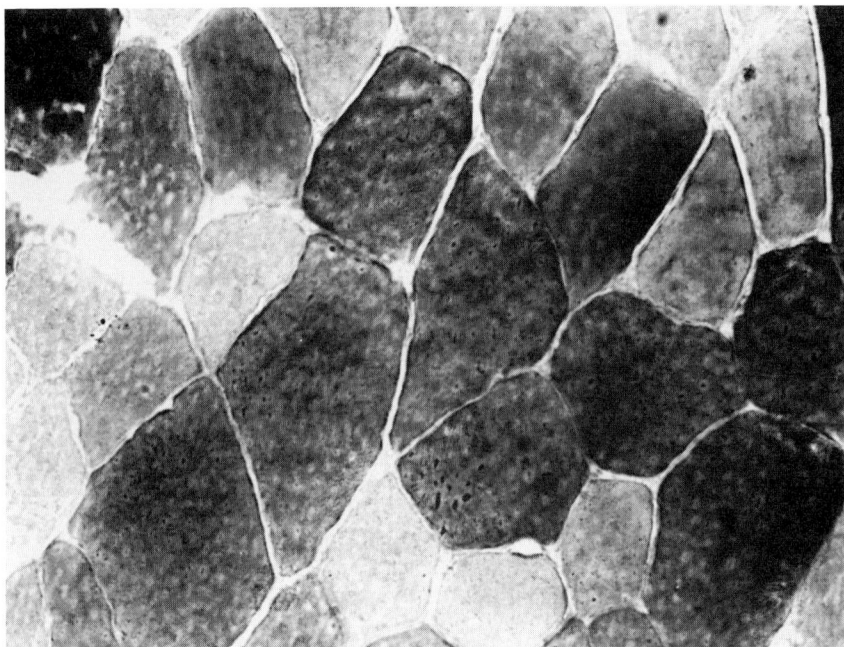

Fig. 2. Unfixed section of human muscle. ATPase method, demonstrating different muscle-fibre types. (× 408)

phosphatase (see p. 29) finalizing with the production of cobalt sulphide. Varying the pH influences ATPase activity.

MYOFIBRILLAR ADENOSINE TRIPHOSPHATASE METHOD (after Brooke and Kaiser 1970)

Sections

Unfixed cryostat.

Preparation of solutions

(A) Veronal acetate

Sodium acetate	1.94 g
Sodium barbitone	2.94 g
Distilled water	100 ml

(B) Pre-incubation buffer, pH 9.4

0.1 M Sodium barbitone (2.062 g/100 ml)	4 ml
0.18 M Calcium chloride $\cdot$ $2H_2O$ (2.648 g/100 ml)	4 ml
Distilled water	12 ml

Adjust to pH 9.4.

(C) Pre-incubation buffer, pH 4.2

Veronal acetate	5 ml
0.1 M HCl	12 ml
Distilled water	6 ml

Adjust to pH 4.2.

(D) Pre-incubation buffer, pH 4.6

Veronal acetate	5 ml
0.1 M HCl	10 ml
Distilled water	8 ml

Adjust to pH 4.6.

(E) Incubating medium

Solution (B)	20 ml
ATP (disodium salt)	50 mg

Adjust pH to 9.4.

Method

1. Pre-incubate three sections at room temperature in solutions: (B) for 15 min; (C) for 5 min; (D) for 5 min
2. Rinse all sections in solution (B) for 30 sec
3. Incubate in medium for 45 min
4. Wash in three changes of 1 per cent calcium chloride for a total of 10 min
5. Transfer to 2 per cent cobalt chloride for 3 min
6. Rinse well in six changes of 0.01 M sodium barbitone
7. Wash well in tap water for 30 sec
8. Place in 1 per cent ammonium sulphide solution for 20–30 sec

   9. Wash in tap water
   10. Mount in glycerine jelly

Results

ATPase activity appears pale, intermediate, or dark depending on fibre type.

6.2. Esterases

Esterases are a sub-group of hydrolytic enzymes which are capable of hydrolyzing esters. By definition, the phosphatases dealt with earlier are strictly esterases, since they hydrolyze phosphate esters. This section is devoted to those enzymes which hydrolyze esters of carboxylic acids. Esterases are divided into non-specific, specific, and lipases; the last may also be regarded as a type of specific esterase. There is considerable overlap between the different types of esterases and many are capable of hydrolyzing the same substrate, making classification difficult. Further subdivision of esterases depends upon the application of inhibitors in enzyme methods.

6.2.1. Inhibitors

Using the inhibitor eserine, the identification of the specific esterases acetyl cholinesterase and cholinesterase is differentiated from non-specific esterases. The hydrolyzing capacity of these *specific esterases* is destroyed while non-specific esterases are not inhibited by eserine. The most useful esterase inhibitors are the organophosphorus compounds such as di-isopropyl fluorophosphate (DFP) and diethyl-*p*-nitrophenyl phosphate (E600). Pearse (1972) sub-divided non-specific esterases into A, B, and C esterases depending on the action of the inhibitors which are summarized in Table 3.

6.2.2. Non-specific esterases

When the substrate is a simple ester such as naphthyl acetate, the hydrolyzing enzyme is termed *non-specific esterase*. Owing to the considerable overlap in activity, more specific esterases such as cholinesterases and lipases are also capable of hydrolyzing simple esters. In addition to the effect of organophosphate inhibitors, non-specific esterases are sub-divided into several different groups, according to the type of ester they hydrolyze most efficiently.

   Aryl esterases   (A type)
   Carboxyl esterases   (B type)
   Acetyl esterases   (C type)

This approximate classification coincides with one based on the effects of organophosphate inhibitors.

Table 3. *Effect of inhibitors on esterase activity.*

Inhibitor	Esterase class A	B	C	Cholinesterase	Lipases
Eserine (10^{-5} M)	UI	UI	UI	I	UI
E600 (10^{-5} M)	UI	I	UI	I	⎰Some are⎱
DFP (10^{-4} M)	UI	I	UI	I	⎱inhibited⎰

UI = Uninhibited; I = inhibited.

METHODS USING NAPHTHYL ACETATE FOR NON-SPECIFIC ESTERASE

A simultaneous coupling method using naphthyl acetate was first described by Nachlas and Seligman (1949). They suggested the use of β-naphthyl acetate as the substrate and diazo blue B as a coupling agent. The azo-dye produced with β-naphthyl acetate is soluble in water, but if α-naphthyl acetate is substituted, this is insoluble, resulting in more precise localization of the enzyme. The method can be used with the diazonium salt Fast blue B as the coupling agent or by the hexazonium pararosaniline technique of Davis and Ornstein (1959). The advantage of using hexazonium pararosaniline coupling is that sections may be dehydrated through alcohol to xylene and mounted in a synthetic mounting medium.

NAPHTHOL ACETATE METHOD USING HEXAZONIUM PARAROSANILINE FOR NON-SPECIFIC ESTERASE (after Davies and Ornstein 1959)

Preparation of solutions

(A) Substrate solution
α-Naphthyl acetate   50 mg
Acetone   5 ml

(B) Buffer solution
Sodium dihydrogen orthophosphate   2.75 g
Distilled water   100 ml

(C) 4 per cent Sodium nitrite solution

(D) Pararosaniline – HC1 stock solution (See p. 34)

(E) Distilled water

Preparation of incubating medium

Solution (A)   0.25 ml
Solution (B)   7.25 ml
Solution (C) ⎱ Mix 0.4 ml of solutions (C) and (D) first   0.8 ml
Soltuion (D) ⎰
Solution (E)   2.5 ml

It is important that equal parts of solutions (C) and (D) are mixed together and allowed to stand for 2 min before adding to the incubating medium. Adjust pH to 7.4 if necessary with solution (B).

Method

1. After suitable fixation, bring sections to water
2. Incubate in above medium at 37 °C for 2–20 min
3. Wash in running water
4. Counterstain in 2 per cent methyl green (chloroform-washed)
5. Wash well in tap water
6. Dehydrate rapidly through fresh alcohol to xylene and mount in DPX.

Results

Esterase appears reddish brown, and nuclei are green.

INDOXYL ACETATE METHOD FOR NON-SPECIFIC ESTERASE (Holt and Withers 1952)

This simultaneous coupling method employs potassium ferricyanide as the coupler. The technique was introduced by Barrnett and Seligman (1951) with the substrate indoxyl acetate. Holt and Withers (1952) used 5-bromo-indoxyl acetate as an alternative substrate giving superior results. In addition to the substrate, the incubating medium consists of potassium ferrocyanide, potassium ferricyanide, calcium chloride, and Tris buffer (pH 7.2). The enzymes in the section hydrolyze 5-bromo-indoxyl acetate to produce 5-bromo-indoxyl which is then oxidized by potassium ferricyanide to an insoluble indigo dye. Potassium ferrocyanide in equimolar solution with ferricyanide prevents over-oxidation of indigo. Calcium chloride acts as an enzyme activator. Sections are mounted in a synthetic medium.

Preparation of incubating medium

4-Chloro-5-bromo-indoxyl-acetate	1 mg
Ethanol	0.1 ml
0.2 M Tris-HCl buffer, pH 7.2	2 ml
Potassium ferricyanide	17 mg
Potassium ferrocyanide	21 mg
Calcium chloride	11 mg
Distilled water	7.9 ml

The substrate is dissolved in the ethanol and the buffer added. The remaining chemicals are dissolved in the distilled water and the solution mixed. It is important that the solution is freshly prepared.

Method

1. After suitable fixation, bring sections to water
2. Incubate in the above medium at 37 °C for 15–60 min
3. Rinse in tap water
4. Counterstain in Mayer's carmalum for 5 min
5. Rinse in tap water
6. Mount in glycerine jelly or (7)
7. Dehydrate through graded alcohols to xylene

8. Mount in DPX

Results

Esterase activity appears blue-green, and nuclei are red.

6.2.3. Specific esterases

Cholinesterases are important specific esterases consisting of *acetyl cholinesterase* ('true') and *cholinesterase* ('pseudo'). Acetyl cholinesterase is capable of hydrolyzing acetyl thiocholine, whereas cholinesterase hydrolyzes other choline esters more rapidly. Both enzymes are capable of hydrolyzing simple esters but cholinesterase is differentiated from non-specific esterase by its capacity to hydrolyze choline esters and by the action of the inhibitor eserine. Acetyl cholinesterase is found in the nervous system, muscle, parafollicular cells of the thyroid, and other endocrine glands.

Another specific esterase is *chloroacetate esterase*. This enzyme is capable of withstanding the effects of standard paraffin processing and the method described by Moloney *et al.* (1960) is reliable. It is preferred for high-resolution work, as the reaction product is insoluble and the section is mounted in a synthetic medium. The technique incorporates a substituted naphthol and is used to demonstrate mast cells and white cells of the myeloid series.

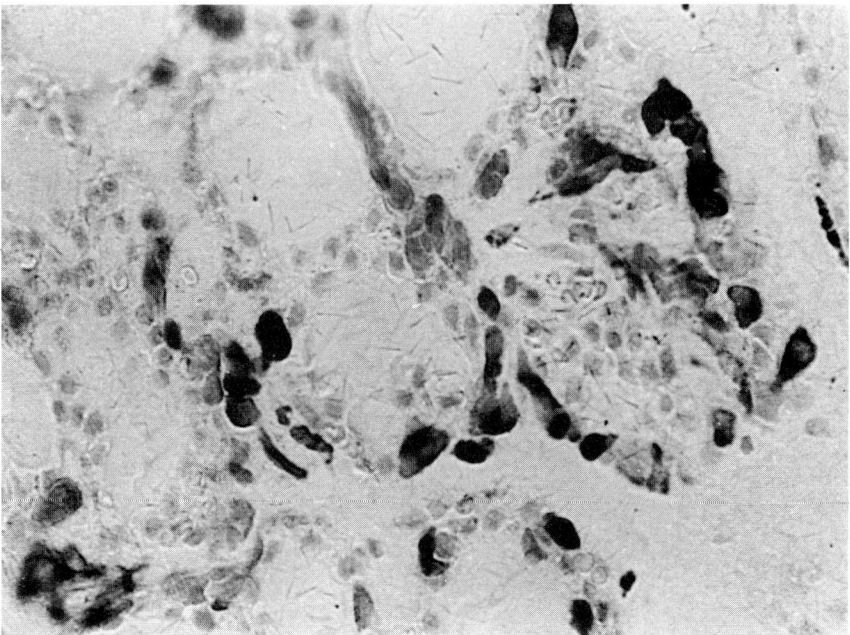

Fig. 3. Section of dog thyroid. Acetylcholinesterase method, demonstrating calcitonin-producing cells. (× 510)

DIRECT-COLOURING THIOCHOLINE METHOD FOR CHOLINESTERASE
(Karnovsky and Roots 1964)

This modification of the original thiocholine technique contains ferricyanide and depends upon copper ions combining with citrate to prevent the formation of copper ferricyanide. Enzymatically-released thiocholine reduces the ferricyanide ion to ferrocyanide which reacts with copper to form insoluble copper ferrocyanide. Both acetyl cholinesterase and cholinesterase are demonstrated, but are differentiated if duplicate sections using the two substrates acetyl and butyryl thiocholine iodine are employed.

Preparation of incubating medium

Acetyl thiocholine iodide	5 mg
0.1 M Acetate buffer pH 6.0	6.5 ml
0.1 M Sodium citrate (2.94 g/100 ml)	0.5 ml
30 mM Copper sulphate (0.58 g/100 ml)	1.0 ml
Distilled water	1.0 ml
5 mM Potassium ferricyanide (0.165 g/100 ml)	1.0 ml

Add in order given, mixing well at each stage.

Method
1. Incubate in above medium at $37\,^{\circ}$C for 15–120 min
2. Wash in distilled water
3. Counterstain in Mayer's haematoxylin
4. Blue in Scott's tap water
5. Rinse in water, alcohol, xylene, and mount in DPX

Results

Cholinesterase activity appears red-brown, and nuclei are blue.

NAPHTHOL CHLOROACETATE METHOD USING HEXAZONIUM PARA-ROSANILINE FOR CHLOROACETATE ESTERASE (Moloney *et al.* 1960)

Preparation of solutions

(A) Substrate solution

Naphthol AS-D chloroacetate	10 mg
N,N-dimethylformamide	1 ml

(B) Buffer solution
0.1M Veronal – HC1 buffer, pH 6.8–7.6

(C) Pararosaniline – HC1 stock solution (See p. 34)

(D) 4 per cent Sodium nitrite

Preparation of incubating medium

Solution (B)	30 ml
Solution (C)	0.4 ml
Solution (D)	0.4 ml

Mix solutions (C) and (D), and allow to stand for 2 min before adding (B). Adjust to pH 6.3 with 1 M HCl, then add solution (A). Filter and use the filtrate as the incubating medium.

Method

1. Bring the sections to water
2. Incubate sections in freshly-prepared filtered incubating medium for 30 min
3. Rinse in water
4. Counterstain nuclei in Mayer's haematoxylin for 2 min
5. Blue in Scott's tap water
6. Rinse in water, dehydrate, clear, and mount in DPX

Results

Chloroacetate esterase activity appears pinkish red, and nuclei are blue.

6.2.4. Lipases

Lipase is a term applied to a group of enzymes having the ability to hydrolyze long-chain esters (more than seven carbon atoms), particularly those containing saturated fatty acids. The enzymes are located in the pancreas and in smaller amounts in the liver and adrenal glands. There is overlap in the demonstration of lipases and non-specific esterases as they hydrolyze the same substrate. Demonstration methods depend on using Tween compounds (which are esters of long-chain fatty acids) and either Sorbitan or Mannitan. Tween-60 indicates that the fatty acid is stearic acid. Using this substrate, fatty acids are produced which combine with calcium ions to form relatively insoluble calcium soaps. These are treated with lead ions and converted by ammonium sulphide to form a dark-brown deposit at the site of enzyme activity.

METAL PRECIPITATION METHOD USING TWEEN FOR LIPASE
(Gomori 1952)

Preparation of solutions

(A) Tris-HCl buffer, pH 7.2

(B) Tween

Either Tween 40, 60, or 80	5 g	
Tris-HCl buffer, pH 7.2	100 ml	
Thymol	1 crystal	

(C) Calcium chloride 200 mg
    distilled water 10 ml

(D) Lead nitrate 1 g
    distilled water 50 ml

Preparation of incubating medium

Solution (A)	9 ml
Solution (B)	0.6 ml
Solution (C)	0.3 ml

Method

1. After suitable fixation, bring sections to water
2. Incubate at 37°C for 2–8 h. If paraffin sections, leave for 24 h
3. Rinse sections in three changes of distilled water
4. Place sections in pre-heated lead nitrate solution (solution D) at 55°C for 10 min
5. Rinse sections in distilled water for 2 min
6. Wash in tap water for 10 min
7. Place sections in 1 per cent ammonium sulphide for 3 min
8. Rinse in distilled water
9. Wash in tap water
10. Counterstain in Mayer's carmalum for 2 min
11. Wash in tap water for 1 min
12. Mount in glycerine jelly

Results

Lipase activity appears yellow to brown-black, and nuclei are red.

Notes

It is advisable to have a control section which is processed through the whole technique, but where the incubating medium lacks Tween. Paraffin sections, according to Gomori, should be fixed in acetone. Formalin-fixed frozen sections work well. Pancreas is a suitable control tissue.

6.3. Miscellaneous

In addition to phosphatases and esterases there are numerous other hydrolytic enzymes including the two disaccharidases (glycosidase) and a peptidase discussed below.

6.3.1. Disaccharidases

Glycosidases are a sub-group of hydrolytic enzymes which hydrolyze glycoside bonds of carbohydrates. A sub-division of glycosidases is disaccharidases. These enzymes occur on the brush border of enterocytes in the small intestine and are essential markers in assessing malabsorption. They are named after the sugar which is hydrolyzed; lactase, trehalase, maltase, iso-maltase, and sucrase. Methods for their demonstration depend on the ability of an enzyme to hydrolyze specific synthetic substrates. The methods for lactase and sucrase are given as these are often the most useful when malabsorption is suspected.

INDIGOGENIC METHOD FOR LACTASE (after Lojda and Kraml 1971)

Sections

Unfixed cryostat.

Preparation of incubating medium

5-Bromo-4-chloro-3-indoxyl-β-D-fucoside	5 mg
N,N-Dimethylformamide	0.5 ml
0.1 M Citric acid phosphate buffer, pH 6.0	10 ml
1.65 per cent Potassium ferricyanide (50 mM/litre)	0.83 ml
2.11 per cent Potassium ferrocyanide (50 mM/litre)	0.83 ml

Add in order given, mixing thoroughly.

Method
1. Incubate in the medium at 37°C for 1–2 h
2. Rinse in distilled water
3. Fix in 4 per cent formaldehyde for 5 min at room temperature
4. Rinse in distilled water
5. Counterstain in Nuclear Fast Red
6. Rinse in water, alcohol, and xylene
7. Mount in Permount

Results

Lactase activity appears turquoise, and nuclei are red.

Note

After incubation, filter the medium and store frozen in a sealed container for future use; at least ten more incubations are possible.

AZO-COUPLING PROCEDURE USING HEXAZONIUM PARAROSANILINE FOR SUCRASE (after Lojda 1965)

Sections

Unfixed cryostat.

Preparation of incubating medium

Solution (1)

(a)	Pararosaniline – HCl stock solution (see p. 34)	0.4 ml
(b)	4 per cent Sodium nitrite	0.4 ml
(c)	0.1 M Citric acid–phosphate buffer, pH 7.0	9.2 ml

Mix solutions (A) and (B), and allow to stand for 2 min before adding (C).

Solution (2)

(d)	6-Bromo-2-naphthyl-α-D-glucoside	3 mg
(e)	N,N-Dimethylformamide	0.5 ml

Then add solution (1) to (2); mix well and adjust to pH 6.5 with 0.1 M NaOH. Filter.

Method
1. Incubate in the medium at room temperature for 1–2 h
2. Rinse in distilled water

3. Fix in 4 per cent formaldehyde for several hours at room temperature
4. Wash in water
5. Counterstain with Mayer's haematoxylin
6. Blue in water
7. Rinse in water, alcohol, and xylene
8. Mount in Permount

Results

Sucrase activity appears orange-red, and nuclei are blue.

6.3.2. Dipeptidyl(amino)peptidase-IV

Another sub-group of hydrolytic enzymes is peptidases which hydrolyze peptide bonds of protein. Dipeptidyl(amino)peptidase-IV (DAP-IV) is an exopeptidase which removes dipeptides from the amino end of the peptide chain, and is found on the brush border of enterocytes in the small intestine and in the cytoplasm of T-lymphocytes. The optimum pH of DAP-IV activity is between 7.2 and 8 and a simultaneous coupling procedure using the diazonium salt Fast Blue BB with the substrate glycyl-proline 4-methoxy naphthylamide is used for the demonstration of the enzyme. Although formalin fixed tissue may be satisfactory, fresh sections are preferred to attain maximum enzyme activity.

AZO-COUPLING PROCEDURE FOR DIPEPTIDY(AMINO)PEPTIDASE-IV
(after Lojda 1981)

Sections

Unfixed cryostat.

Preparation of incubating medium

Solution (1)

(A) Fast Blue BB salt	5 mg	
(B) Distilled water	1 ml	
(C) 0.1 M phosphate buffer, pH 7.2	4 ml	

Mix (A) and (B), then add (C).

Solution (2)

(D) Glycyl-proline-4-methoxynaphthylamide	2 mg	
(E) N,N-dimethylformamide	2 drops	

Then add solution (1) to (2). Filter before use.

Method

1. Fix in equal parts of chloroform and acetone at $4°C$ for 1 min
2. Incubate in the medium at $37°C$ for 30 min
3. Wash in distilled water
4. Counterstain in Mayer's haematoxylin

5. Blue in water
6. Mount in glycerine jelly

Results

DAP-IV activity appears red, and nuclei are blue.

7

Transferases and oxidative enzymes

7.1. Transferases

Transferases are enzymes which catalyse the transfer of certain groups from one compound to another. Few transferases are found in animal tissue, but the most important is phosphorylase. *In vivo*, phosphorylase is a cytoplasmic enzyme concerned with the breakdown of glycogen by rupturing α-1,4-glycosidic bonds in the presence of phosphate to produce glucose-l-phosphate. However, the histochemical technique relies on the synthesis of polysaccharide chains from glucose-l-phosphate. A polysaccharide primer is necessary for the reaction to proceed. The amount of phosphorylase present determines the length of the unbranched chains, which are stained with iodine. The colour varies according to the chain length, from yellow through to brown, lavender, and purple to intense blue-black for the longest chains.

TECHNIQUE FOR MYOPHOSPHORYLASE (Filipe and Lake 1983)

Sections
Unfixed cryostat.

Preparation of incubation medium

0.1 M Acetate buffer, pH 5.9	10 ml
0.1 M Magnesium chloride	1 ml
Glucose-l-phosphate (dipotassium salt)	100 mg
Glycogen (oyster or rabbit)	2 mg
Adenosine 5-triphosphate	5 mg
Sodium fluoride	180 mg
Ethanol	2 ml
Polyvinylpyrolidine (PVP)	0.9 g

Store aliquots at -20°C.

Method
1. Incubate in the medium at 37°C for 1 h
2. Rinse in 40 per cent ethanol for 5 sec
3. Air-dry rapidly
4. Fix for in ethanol for 3 min
5. Air-dry rapidly
6. Stain in water: Lugol's iodine (10:1) for 5 min
7. Mount in glycerine jelly: Lugol's iodine (9:1)

Results

Phosphorylase activity appears yellow–blue black (see above).

Note

Fading occurs fairly rapidly, but staining can be renewed by repeating steps 6 and 7.

7.2. Oxidative enzymes

Oxidative enzymes have the ability to catalyse the oxidation of substrates. The way in which this oxidation is achieved gives rise to the following sub-groups: *Oxidases; peroxidases; dehydrogenases* (including *diaphorases*).

Oxidation is usually achieved by removal of hydrogen. Each reaction is also accompanied by a simultaneous reduction reaction, and the overall process is more correctly expressed as a transfer of electrons.

Preparation of the tissue for the demonstration of oxidative enzymes is critical as many enzymes are unable to withstand the effects of fixation. Consequently, fresh unfixed frozen sections are usually employed. Most enzyme-demonstration methods available are for oxidases and dehydrogenases with many ultilizing tetrazolium salts.

7.2.1. Oxidases

Oxidases are enzymes which catalyze the oxidation of substrate in the presence of oxygen. It is usual that hydrogen is removed from the substrate, which then combines with oxygen to form water, rather than oxygen attaching itself directly to the substrate. Oxidases are distinguished from dehydrogenases by the fact that they are unable to function anaerobically.

$$AH_2 + 1/2\,O_2 \longrightarrow A + H_2O$$
(Substrate)   (Oxidized substrate)

7.2.1.1. Tyrosinase (DOPA oxidase)

Tyrosinase (or *DOPA oxidase)* catalyzes the oxidation of tyrosine to dihydroxphenylalanine (DOPA) and after further oxidation to the pigment *melanin*. The method given is used to demonstrate cells capable of producing melanin, including those in malignant melanomas.

TYROSINASE-DOPA REACTION (Okun *et al.* 1969)

Fixation and sections

Formalin-fixed frozen or cryostat sections; unfixed cryostat sections.

Preparation of solutions

Control incubating solution (A)
   0.1 M Phosphate buffer, pH 7.4   10 ml

Test incubating solution (B)

L-Tyrosine	2 mg
D,L-DOPA	0.2 mg
0.1 M Phosphate buffer, pH 7.4	10 ml

Control incubating solution (C)

D,L-DOPA	0.2 mg
0.1 M Phosphate buffer, pH 7.4	10 ml

Control incubating solution (D)

L-Tyrosine	2 mg
D,L-DOPA	0.2 mg
0.1 M Phosphate buffer, pH 7.4	10 ml
Sodium diethyldithiocarbamate	1 mg

Method
1. Label four serial or near-serial sections A, B, C, and D
2. Place the sections in the appropriate solutions (i.e. slide A in solution (A), etc.) for 3 h at 37°C
3. Rinse in phosphate buffer, pH 7.4, for 2 min
4. Wash in distilled water for 2 min
5. Dehydrate through graded alcohols to xylene and mount in DPX

Results

In section A, preformed pigment only is seen. In section B a black pigment, absent from other sections, indicates melanin synthesis. (Only sections B and C demonstrate enzyme activity.) In section C there is little induced pigment, while in section D no new pigment is seen.

7.2.1.2. Monoamine oxidase

Monoamine oxidase is found in several tissues including the adrenal glands, liver, and heart muscle. The method below involves the oxidation of tryptamine by the enzyme, with the resulting product reducing the tetrazolium salt tetranitro-blue tetrazolium (TNBT), to produce a formazan deposit.

MONOAMINE OXIDASE TETRAZOLIUM METHOD (Glenner *et al.* 1957)

Sections
Unfixed cryostat.

Preparation of incubating medium

Tryptamine – HC1	25 mg
Sodium sulphate	4 mg
TNBT	5 mg
0.1 M Phosphate buffer, pH 7.6	5 ml
Distilled water	15 ml

Method
1. Place sections in incubating medium at 37°C for 45 min
2. Wash in running tap water for 2 min
3. Place sections in 10 per cent formol saline for 30 min
4. Wash well in tap water for 2 min
5. Mount in glycerine jelly

Results

Monoamine oxidase activity appears blue-black.

7.2.1.3. Cytochrome oxidase

Cytochrome oxidase is found in many tissues, being particularly plentiful in liver, kidney, and muscle. Cytochrome oxidase (also known as *cytochrome a_3*), along with other cytochromes, is involved in the main oxidation pathways carried out in mitochondria.

METAL-CHELATION METHOD FOR CYTOCHROME OXIDASE (Burstone 1959)

Sections

Unfixed cryostat.

Preparation of incubating medium

Solution (A)

1-Hydroxyl-2-acetonaphthone	10 mg
N-phenyl-*p*-phenylenediamine	10 mg
Absolute alcohol	0.5 ml
Distilled water	35 ml
0.2 M Tris buffer, pH 7.4	15 ml

The first two reagents are dissolved in the absolute alcohol, and then the distilled water and buffer are added.

Solution (B)

Cobalt acetate	5 g
Formaldehyde (40 per cent w/v aqueous solution)	5 ml
Distilled water	45 ml

Method
1. Place sections into incubating solution (A) for 15 min to 2 h
2. Transfer sections directly to 10 per cent cobalt acetate (B), for 1 h
3. Wash in distilled water
4. Mount in glycerine jelly

Results

Cytochrome oxidase activity appears blue-black.

7.2.2. Peroxidases

Peroxidases catalyze the oxidation of the substrate by removing hydrogen, which combines with hydrogen peroxide to form water.

$$AH_2 + H_2O_2 \longrightarrow A + 2H_2O$$
(Substrate)   (Oxidized substrate)

Peroxidases are present in red and white blood cells. The enzymes are resistant to the effects of aldehyde fixation and are demonstrated using 3,3'-diaminobenzidine (DAB) as the substrate. *DAB is a possible carcinogen and the user should wear gloves whilst handling it.*

PEROXIDASE TECHNIQUE USING DAB (Graham and Karnovsky 1966)

Preparation of incubating medium

3,3'-Diaminobenzidine tetrahydrochloride	5 mg
Tris-HCl buffer, pH 7.6	10 ml
1 per cent Hydrogen peroxide	1.2 ml

Method
1. Rinse sections in distilled water
2. Incubate in the above medium for 5 min
3. Wash well in distilled water
4. Counterstain 2 per cent methyl green (chloroform-washed)
5. Wash in water
6. Dehydrate, clear, and mount in DPX

Results

Peroxidase appears brown, and nuclei are green.

7.2.3. Dehydrogenases

Dehydrogenases are enzymes which have the ability to remove hydrogen from the substrate and transfer it to another substance. The substance which acts as the hydrogen acceptor is either a coenzyme, NAD or NADP, or a flavoprotein. Oxidation of a substrate (A) by dehydrogenases may be considered as:

$$AH + NAD \longrightarrow A + NADH$$
(Reduced)   (Oxidized)

7.2.3.1. Diaphorases

Diaphorases are dehydrogenases which catalyse the dehydrogenation of the reduced forms of NAD and NADP, i.e. they catalyze the reactions:

$$NADH \longrightarrow NAD + H^+$$

$$\text{NADPH} \longrightarrow \text{NADP} + \text{H}^+$$
$$\text{(Reduced)} \qquad\qquad \text{(Oxidized)}$$

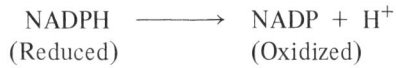

Dehydrogenases are mitochondrial enzymes and are inactivated by standard fixation techniques. Fresh unfixed cryostat sections are required.

DEMONSTRATION OF DEHYDROGENASES

The demonstration of dehydrogenases and diaphorases relies on the reduction of tetrazolium salts by hydrogen ions (released during oxidation) to produce formazans. Three tetrazolium salts commonly used are MTT, NBT and TNBT, any of which may be employed for the enzymes listed below. If MTT is used, the formazan chelates with cobalt ions to produce the final coloured deposit. If NBT or

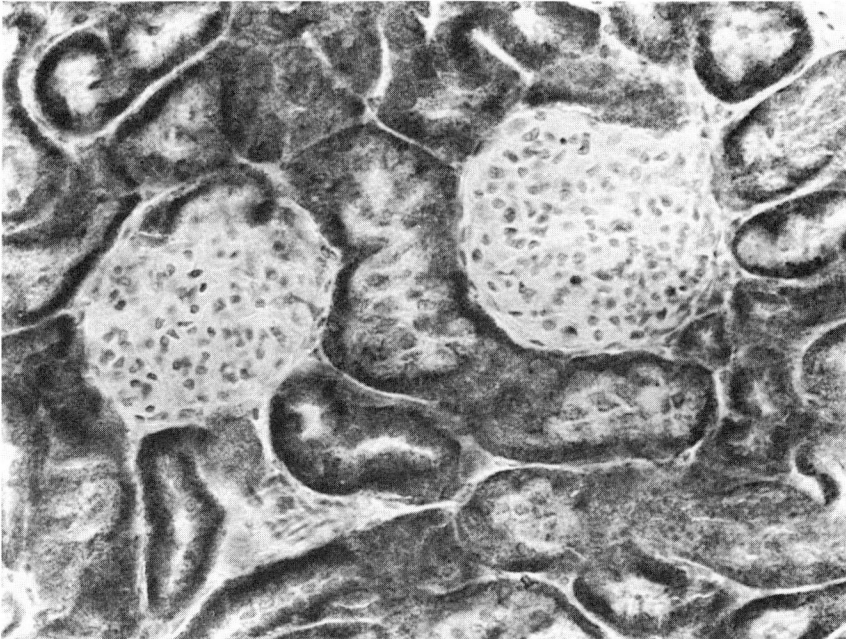

Fig. 4. Unfixed section of rat kidney, demonstrating succinic dehydrogenase with MTT. ($\times$ 340)

TNBT is chosen, then cobalt is not required. A method is given below to illustrate the techniques of dehydrogenase histochemistry. The same method can be applied for the demonstration of many dehydrogenases, by applying a specific substrate with the addition of a tetrazolium salt solution. Some enzymes also require a co-enzyme. After incubation, the sections are transferred to formol saline which fixes the tissue.

Preparation of tetrazolium stock solution

MTT (1 mg/ml distilled water) (For NBT or TNBT, see below)	2.5 ml
Tris-HCl buffer, pH 7.4	2.5 ml
0.5 M Cobalt chloride	0.5 ml
0.05 M Magnesium chloride	1.0 ml
Distilled water	2.5 ml

Check the pH and adjust to 7.0 using either stock buffer or 1 M HCl. The stock solution is kept frozen and is stable for many months in this form. Coenzymes are added just before use. Check the pH is 7.0 (For NBT or TNBT solution, use a concentration of 4 mg/ml of distilled water, omit the cobalt chloride and add 3 ml distilled water.) For substrate stock solutions and incubating medium, see Tables 4 and 5 (on p. 55).

Method
1. Cover sections with the incubating medium at 37 °C for 30 min to 1 h
2. Transfer sections to formol saline for 10–15 min
3. Wash well in tap water for 2 min
4. Counterstain in 2 per cent methyl green (chloroform-washed)
5. Rinse in tap water
6. Mount in glycerine jelly
7. If NBT or TNBT has been used as the tetrazolium salt, dehydrate through alcohol to xylene and mount in DPX.

Results

Dehydrogenase appears black with MTT, or purple with NBT and TNBT; nuclei are green.

Notes

Mayer's carmalum is a suitable counterstain with NBT and TNBT. Avoid dehydration with acetone if NBT is used as the formazan will be removed.

Table 4. *Preparation of stock substrate solutions for demonstration of dehydrogenases with pH adjusted to 7.0.*

Substrate	Chemical	Conc. (M)	Amount of substrate	Vol. of water (ml)	Neutralize with:	Distilled water (ml)[a]
Succinate	Sodium succinate	2.5	6.75 g	8	1 M HCl	10
Malate	Sodium hydrogen malate	1	1.55 g	8	40% NaOH	10
Glucose-6-phosphate	Glucose-6-phosphate (disodium salt)	1	0.30 g	0.8	1 M HCl	1.0
Isocitrate	D,L-isocitric acid (trisodium salt)	1	0.27 g	0.8	1 M HCl	1.0
Glutamate	L-glutamic acid (Na salt) monohydrate	1	1.87 g	8	1 M HCl	10
Lactate	Sodium D,L-lactate	1	1.25 ml	8.75	-	10

[a] Make up to find volume with distilled water.

Table 5. *Working solutions for demonstration of diaphorases and dehydrogenases*[a]

Enzyme	Vol. of substrate stock soln. (ml)	Vol. of distilled water (ml)	Coenzyme (2 mg)
NAD diaphorase	-	0.1	NADH
NADP diaphorase	-	0.1	NADPH
Succinate dehydrogenase	0.1	-	-
Malate dehydrogenase	0.1	-	NAD
Glucose-6-phosphate dehydrogenase	0.1	-	NADP
Isocitrate dehydrogenase	0.1	-	NAD
Glutamate dehydrogenase	0.1	-	NAD
Lactate dehydrogenase	0.1	-	NAD

[a] Add 0.9 ml of stock tetrazolium solution (p. 54) in all cases.

8

Diagnostic applications

8.1. Introduction

Enzymes are present in all tissues. Although they are essential to promote chemical reactions in biological systems, their diagnostic significance is more restricted. Their diagnostic applications in histopathology will be briefly discussed under the following headings: *Skeletal muscle fibre typing; lymphoid and myeloid cells; malabsorption; Hirschsprung's disease; miscellaneous.*

Only an outline of each is given in this handbook; for further reading, the book by Filipe and Lake (1983) is recommended.

8.2. Skeletal muscle fibre typing

The application of enzyme histochemical methods to *skeletal muscle* has revealed several different fibre types, which on examining their proportion and size can assist in establishing a diagnosis. Classification of muscle disease has become complex (see Weller 1984) but muscle fibres may be broadly divided into Type 1 and Type 2 depending on the level of adenosine triphosphatase (ATPase) present. Incubation at pH 9.4 for this enzyme shows Type 1 fibres to exhibit low ATPase activity in contrast to Type 2 fibres which show high ATPase activity. Demonstration of a mitochondrial enzyme such as succinate dehydrogenase reveals the opposite staining patterns, i.e. Type 1 fibres stain strongly but Type 2 fibres are weakly stained.

If ATPase staining is applied after pre-incubating at pH 4.2 or 4.6 in buffer, Type 2 fibres may be further divided into sub-types 2A, and 2B, and 2C. Consequently, for the differentiation of fibre types in skeletal muscle, staining for ATPase at pH 4.2, 4.6, and 9.4 is employed. In addition, staining for phosphorylase is useful as Type 1 fibres exhibit low activity and Type 2 fibres high activity. Phosphorylase in McArdle's disease is deficient in muscle fibres although its activity in smooth muscle of blood vessels is unaffected, which serves as a useful inherent control.

The results of these staining methods applied to normal human skeletal muscle are summarized in Table 6. The application of these histochemical methods requires fresh unfixed cryostat sections. It is essential that muscle is snap-frozen rapidly using isopentane super-cooled with liquid nitrogen, as described in Chapter 4, to avoid ice-crystal artefact. The tissue is cut into blocks no more than $0.5\,cm^3$ and orientated using a dissecting microscope so the fibres are sectioned transversely. Compressed areas of muscle resulting from the use of clamps should not be used for examination.

Table 6. *Histochemical fibre types in human skeletal muscle based on enzymic properties.*[a]

Fibre type	Succinate dehydrogenase	ATPase activity at pH:			Phosphorylase
		4.2	4.6	9.4	
Type 1	+++	+++	+++	+	+
Type 2A	++	—	—	+++	+++
Type 2B	+	—	+++	+++	+++
Type 2C[b]	++	++	+++	+++	+++

[a] Based on Dubowitz 1985.
[b] Normal adult muscle contains few Type 2c fibres.

8.3. Lymphoid and myeloid cells

Several enzymes applicable to smears have already been described (p. 22) as suitable markers for *lymphoid cells*. In addition to specific cells exhibiting enzyme activity, a further criteria of differentiation is their staining pattern, i.e. focal or diffuse. Chloroacetate esterase is perhaps the most useful enzyme to distinguish lymphoid and *myeloid cells* as the former are not demonstrated. Mast cells are stained but are easily recognized by their morphology and granularity. Another advantage of chloroacetate esterase is its ability to withstand the effects of fixation and processing, allowing its demonstration in both paraffin wax and resin (e.g. GMA) sections. In many laboratories bone-marrow trephines are routinely embedded in resin.

8.4. Malabsorption

To obtain a definitive diagnosis on gastrointestinal biopsies where *malabsorption* is suspected, the morphological information provided by paraffin sections is insufficient. A more accurate diagnosis is derived from an enzyme histochemical profile using frozen sections, by incubating for alkaline phosphatase, acid phosphatase, and disaccharidases, with the last being of fundamental importance. Disaccharidases which may be demonstrated include lactase, trehalase, and sucrase and reflect changes within the tissue; lactase is the most sensitive and sucrase the most resistant to enterocyte injury. These enzymes and alkaline phosphatase are found on the brush border of villi, whereas acid phosphatase is present in the lysosomes of enterocytes and also within macrophages in the lamina propria. Malabsorption can only be properly assessed using jejunum which is orientated so the villi are sectioned longitudinally.

8.5. Hirschsprung's disease

In *Hirschsprung's disease* there is an absence of ganglion cells and a marked increase in the number and thickening of non-argyrophilic nerve fibres in the lower part of the colon and rectum. These nerve fibres which are found between the muscle bundles in the muscularis mucosae and in the lamina propria, contain cholinesterase and are demonstrated using an acetylcholinesterase method as previously described. Since the investigation is carried out whilst the patient is still on the operating table, speed is essential, and consequently frozen sections are utilized. The demonstration of ganglion cells using a technique for non-specific esterase may be of more immediate diagnostic value than the longer time required for cholinesterase staining of nerve fibres. Several rectal and colonic biopsies may be required to ascertain that only the abnormal aganglionic segment is removed, before anastomosis between the normal ganglionic bowel ends can be performed.

8.6. Miscellaneous

There are many situations where enzyme histochemistry may be of diagnostic value. For example, the demonstration of acid phosphatase in the identification of *prostatic carcinoma* and esterase activity for *thyroid carcinoma*. These are both useful in diagnosing a primary tumour from a metastasis. In cervical carcinoma the amount of esterase increases with the severity of the tumour. In bone, acid and alkaline phosphatase may be demonstrated in osteoclasts and active osteoblasts, respectively.

Appendix 1

List of methods and control tissues

Enzyme	Page	Control tissue
Acid phosphatase	32	Kidney, liver, prostate
Alkaline phosphatase	29	Kidney, small intestine
Chloroacetate esterase	42	Bone marrow, skin (mast cells)
Cholinesterase	42	Liver, small intestine, brain
Cytochrome oxidase	51	Liver, muscle, kidney (mitochondria)
Dehydrogenases, e.g. succinate dehydrogenase	53	Liver, muscle, kidney (mitochondria)
Dipeptidyl(amino)peptidase-IV	46	Tonsil, small intestine
Lactase	44	Jejunum
Lipase	43	Pancreas
Monoamine oxidase	50	Adrenal, liver, heart
Myofibrillar adenosine triphosphatase	37	Skeletal muscle
Myophosphorylase	48	Skeletal muscle
Non-specific esterase	39	Liver, kidney
Peroxidase	52	Granulocytes
Tyrosinase	49	Skin
Sucrase	45	Jejunum

Appendix 2

Preparation of solutions

4 per cent Neutral formol calcium

40 per cent formaldehyde (analytical grade)	10 ml
Distilled water	90 ml
Calcium chloride (until pH 7.0) approx.	1.1 g

Formalin acetone buffer

Disodium hydrogen phosphate	20 mg
Potassium dihydrogen phosphate	100 mg
Distilled water	30 ml
Acetone	45 ml
Formalin	25 ml

Adjust to pH 6.6.

Formaldehyde–gelatine adhesive for frozen sections

1 per cent gelatine	25 ml
2 per cent formalin	25 ml

Heat gently to dissolve.

Gum sucrose

Gum acacia	1 g
Sucrose	30 g
Distilled water	100 ml

Heat gently to dissolve, and store at 4°C.

Lugol's iodine

Iodine	1 g
Potassium iodide	2 g
Distilled water	100 ml

Dissolve the potassium iodide before adding the iodine; gently heat if necessary.

Appendix 3

Buffers

1. Acetate buffer 0.1 M (Walpole)

Stock A
  0.2 M Acetic acid (MW = 60.05)
  1.2 ml Glacial acetic acid in 100 ml of distilled water

Stock B
  0.2 M Sodium acetate
  1.64 g Anhydrous sodium acetate (MW = 82) or 2.75 g sodium acetate trihydrate (MW = 136) in 100 ml of distilled water

pH	A (ml)	B (ml)
4.0	41.0	9.0
4.2	36.8	13.2
4.4	30.5	19.5
4.6	25.5	24.5
4.8	20.0	30.0
5.0	14.8	35.2
5.2	10.5	39.5
5.4	8.8	41.2
5.6	4.8	45.2

Methods requiring a pH higher than 5.6 (5.9 and 6.0), add to 46 ml of stock B, 0.2 M acetic acid (stock A) until correct pH is reached and make up to 50 ml with solution B. Finally, add 50 ml of distilled water.

2. 0.1 M Cacodylate buffer, pH 7.4

    Sodium cacodylate (MW = 214)   2.14 g
    Distilled water   80 ml

Adjust the pH with 0.1 M HCl and make up to a final volume of 100 ml.

3. Citric acid phosphate buffer 0.1 M (McIlvaine)

Stock A
    0.1 M Citric acid (MW = 210.0)
    2.1 g Citric acid in 100 ml of distilled water

Stock B
    0.2 M Disodium hydrogen orthophosphate (MW = 142.0)
    2.83 g Disodium hydrogen orthophosphate in 100 ml of distilled water

pH	A (ml)	B (ml)
5.0	48.5	51.5
5.2	46.4	53.6
5.4	44.3	55.7
5.6	42.0	58.0
5.8	39.6	60.4
6.0	36.9	63.1
6.2	33.9	66.1
6.4	30.8	69.2
6.6	27.3	72.7
6.8	22.8	77.2
7.0	17.7	82.3
7.2	13.1	86.9
7.4	9.2	90.8
7.6	6.4	93.6
7.8	4.3	95.7
8.0	2.8	97.2

4. Phosphate Buffer 0.1 M

Stock A
 0.2 M Sodium dihydrogen orthophosphate (MW = 156)
 3.12 g Sodium dihydrogen orthophosphate in 100 ml of distilled water

Stock B
 0.2 M Disodium hydrogen orthophosphate (MW = 142)
 2.83 g Disodium hydrogen orthophosphate in 100 ml of distilled water

The buffer is composed of x ml A + y ml B made up to 100 ml with distilled water.

pH	A (x ml)	B (y ml)
6.0	43.8	6.2
6.2	40.7	9.3
6.4	36.7	13.3
6.6	31.2	18.8
6.8	25.5	24.5
7.0	19.5	30.5
7.2	14.0	36.0
7.4	9.5	40.5
7.6	6.5	43.5
7.8	4.2	45.8
8.0	2.6	47.4

5. Tris-HCl buffer 0.2 M

Stock A
 0.2 M Tris (MW = 121.0)
 2.42 g Tris(hydroxymethyl)methylamine in 100 ml of distilled water

Stock B
 0.2 M HCl (MW = 36.46)
 1.7 ml Hydrochloric acid in 100 ml of distilled water

The buffer is composed of 25 ml A + x ml B, made up to 100 ml with distilled water.

pH	B (x ml)
7.2	22.1
7.4	20.7
7.6	19.2
7.8	16.3
8.0	13.4
8.2	11.0
8.4	8.3
8.6	6.1
8.8	4.1
9.0	2.5

6. Veronal–HCl, buffer 0.1 M (Michaelis)

Stock A
   0.1 M Sodium veronal (MW = 206.18)
   2.062 g Sodium barbitone in 100 ml of distilled water

Stock B
   0.1 M Hydrochloric acid (MW = 36.46)
   0.85 ml Hydrochloric acid in 100 ml of distilled water

pH	A (x ml)	B (y ml)
6.8	20.9	19.1
7.0	21.4	18.6
7.2	22.2	17.8
7.4	23.2	16.8
7.6	24.6	15.4
7.8	26.5	13.5
8.0	28.6	11.4

References

Ashley, C.A. and Feder, N. (1966). Glycol methacrylate in histopathology. *Arch. Path.* **81**, 391.

Bancroft, J.D. (1966). Observations on the effect on histochemical reactions of different processing methods. *J. Med. Lab. Technol.* **23**, 105.

——(1975). *Histochemical technique*, (2nd edn). Butterworths, London.

—— and Stevens, A. (1982). Theory and practice of histological techniques, (2nd edn). Churchill-Livingstone, Edinburgh.

Barka, T. (1960). A simple azo-dye method for histochemical demonstration of acid phosphatases. *Nature* **187**, 248.

Barrnett, R.J. and Seligman, A.M. (1951). Histochemical demonstration of esterases by production of indigo. *Science* **114** 579.

Beckstead, J.H. and Bainton, D.F. (1980). Enzyme histochemistry on bone marrow biopsies: reactions useful in the differential diagnosis of leukemia and lymphoma applied to 2-micron plastic sections. *Blood* **55**, 386.

—— (1983). The evaluation of human lymph nodes, using plastic sections and enzyme histochemistry. *Am. J. Clin. Pathol.* **80**, 131.

Boadle, M.C. and Bloom, F.E. (1969). A method for the fine structural localisation of monoamine oxidase. *J. Histochem. Cytochem.* **17**, 331.

Briggs, R.T., Drath, D.B., Karnovsky, M.L. and Karnovsky, M.J. (1975). Localization of NADH oxidase on the surface of human polymorphonuclear leucocytes by a new cytochemical method. *J. Cell. Biol.* **67**, 566.

Brooke, M.H. and Kaiser, K.K. (1970). Three 'myosin adenosine triphosphatase' systems: the nature of their pH lability and sulfhydryl dependence. *J. Histochem. Cytochem.* **18**, 670.

Burnett, R. (1982). The use of histochemical techniques on 1 micron methacrylate sections of kidney in the study of cephaloridine nephrotoxicity in rats. In *Nephrotoxicity: assessment and pathogenesis*, (eds P.H. Bach, F.W. Bonner, J.W. Bridges and E.A. Locks), p. 98. John Wiley, Chichester.

Burstone, M.S. (1958a). Histochemical comparison of naphthol AS-phosphates for the demonstration of phosphatases. *J. Nat. Cancer Inst.* **20**, 601.

—— (1985b). Histochemical demonstration of acid phosphatase with naphthol AS-phosphates. *J. Nat. Cancer Inst.* **21**, 523.

—— (1959). New histochemical techniques for the demonstration of tissue oxidase (cytochrome oxidase). *J. Histochem. Cytochem.* **7**, 112.

—— (1962). *Enzyme histochemistry and its application to the study of neoplasms.* Academic Press, New York.

Catovsky, D., Cherchi, M., Greaves, M.F., Janossy, G., Pain, C. and Kay, H.E.M. (1978). Acid phosphatase reaction in acute lymphoblastic leukaemia. *Lancet* 1, (April), 749.

Chang, J.P. and Hori, S.H. (1962). Survival of enzymes in section frozen-substituted tissues. *J. Histochem. Cytochem.* **10**, 592.

Chilosi, M., Pizzolo, G., Menestrina, F., Iannucci, A.M., Bonetti, F. and Fiore-Donati, L. (1981). Enzyme histochemistry and normal and pathologic paraffin-embedded lymphoid tissues. *Am. J. Clin. Pathol.* **76**, 729.

Christensen, A.K. (1971). Frozen thin sections of fresh tissue for electron microscopy, with a description of pancreas and liver. *J. Cell. Biol.* **51**, 772.

Davis, B.J. and Ornstein, L. (1959). High resolution enzyme localisation with a new diazo reagent hexazonium pararosaniline. *J. Histochem. Cytochem.* **7**, 297.

Dawson, I.M.P. (1972). Fixation: What should the pathologist do? *Histochem. J.* **4**, 381.

Dixon, M. and Webb, E.C. (1979). *Enzymes*, (3rd edn). Longman, London.

Dubowtiz, V. (1985). *Muscle biopsy: a practical approach*, (2nd edn). Baillière-Tindall, London.

Feder, N. and Sidman, R.L. (1958). Histological fixation by a modified freeze substitution method. *J. Histochem. Cytochem.* **6**, 401.

Filipe, M.I. and Lake, B.D. (1983). *Histochemistry in pathology*. Churchill-Livingstone, Edinburgh.

Fujimori, T., Mochino, T., Miura, M. and Katayama, I. (1981). Enzyme histochemistry on paraffin embedded tissue sections. *Stain Technol.* **56**, 335.

——, Inomata, K. and Ogawa, K. (1982). A cerium method for the ultracytochemical localization of monoamine oxidase activity. *Histochem. J.* **14**, 87.

Gibbons, I.R. (1959). An embedding resin miscible with water for electron microscopy. *Nature* **184**, 375.

Glenner, G.G., Burtner, H.J. and Brown, G.W. (1957). The histochemical demonstration of monoamine oxidase activity by tetrazolium salts. *J. Histochem. Cytochem.* **5**, 592.

Golberg, A.F. and Barka, T. (1962). Acid phosphatase activity in human blood cells. *Nature* **195**, 297.

Gomori, G. (1941). Distribution of acid phosphatase in the tissues under normal and under pathological conditions. *Arch. Path.* **32**, 189.

—— (1951). Alkaline phosphatase of cell nuclei. *J. Lab. Clin. Med.* **37**, 526.

—— (1952). *Microscopic histochemistry: principles and practice*. University of Chicago Press, Chicago.

Graham, R.C. and Karnovsky, M.J. (1966). The early stages of absorption of injected horseradish peroxidase in the proximal tubules of mouse kidney: ultrastructural cytochemistry by a new technique. *J. Histochem. Cytochem.* **14**, 291.

Grogg, E. and Pearse, A.G.E. (1952). A critical study of the histochemical techniques for acid phosphatase, with a description of an azo-dye method. *J. Path. Bact.* **64**, 627.

Hanker, J.S., Anderson, W.E. and Bloom, F.E. (1971). Osmiophilic polymer generation. Catalysis by transition metal compounds in ultrastructural cytochemistry. *Science* **175**, 991.

Hayat, M.A. (1973, 1974a, 1975, and 1977). *Electron microscopy of enzymes*. Vols 1, 2, 3, 4, and 5 respectively. Van Nostrand Reinhold, New York.

Hess, R., Scarpelli, D.G. and Pearse, A.G.E. (1958). Cytochemical localisation of pyridine nucleotide linked dehydrogenase. *Nature* **181**, 1531.

Higgy, K.E., Burns, G.F. and Hayhoe, F.G.J. (1977). Discrimination of B, T and null lymphocytes by esterase cytochemistry. *Scand. J. Haematol.* **18**, 437.

Higuchi, S., Suga, M., Dannenberg, Jr., A.M. and Scholfield, B.H. (1979). Histochemical demonstration of enzyme activities in plastic and paraffin embedded tissue sections. *Stain Technol.* **54**, 5.

Holt, S.J. and Withers, R.F.J. (1952). Cytochemical localisation of esterases using indoxyl derivatives. *Nature* **170**, 1012.

Holt, S.J., Hobbiger, E.L. and Pawan, G.L.S. (1960). Preservation of integrity of rat tissues for cytochemical staining purposes. *J. Biophys. Biochem. Cytol.* **7**, 383.

Holt, S.J. and Hicks, R.M. (1961). Studies on formalin fixation for electron microscopy and cytochemical staining purposes. *J. Biophys. Biochem. Cytol.* **11**, 31.

Horton, A.W., Dockery, N., Sillence, D. and Rimoin, D.L. (1980). An embedding method for histochemical studies of undecalcified skeletal growth plate. *Stain Technol.* **55**, 19.

Hulstaert, C.D., Kalichanaran, D. and Hardonk, M.J. (1983). Cytochemical demonstration of phosphatases in the rat liver by a cerium-based method in combination with osmium tetroxide and potassium ferrocyanide postfixation. *Histochemistry* **78**, 71.

Karnovsky, M.J. and Roots, L. (1964). A 'direct-coloring' thiocholine method for cholinesterase. *J. Histochem. Cytochem.* **12**, 219.

Lewis, P.R. and Knight, D.P. (1977). Staining methods for sectioned material. In *Practical methods in electron microscopy*, (ed. A.M. Glauert), Vol. 5, part 1. North-Holland, Amsterdam.

Li, C.Y., Yam, L.T. and Lam, K.W. (1970). Acid phosphatase isoenzyme in human leucocytes in normal and pathologic conditions. *J. Histochem. Cytochem.* **18**, 473.

Lojda, Z. (1965). Some remarks concerning the histochemical detection of disaccharidases and glucosidases. *Histochemie* **5**, 339.

—— and Kraml, J. (1971). Indigogenic methods for glycosidases. III. An improved method with 4-Cl, 5-Br, 3-indoxyl-β-D-fucoside and its application in studies of enzymes in the intestine, kidney and other tissues. *Histochemie* **25**, 195.

—— (1977). Studies on glycyl-proline naphthylamidase. I. Lymphocytes. *Histochemistry* **54**, 299.

—— (1981). Protinases in pathology, Usefulness of histochemical methods. *J. Histochem. Cytochem.* **29**, 481.

Menton, M.L., Junge, J. and Green, M.H. (1944). A coupling histochemical azo-dye test for alkaline phosphatase in the kidney. *J. Biol. Chem.* **153**, 471.

Meryman, H.T. (1960). Principles of freeze drying. *Annals of the New York Academy of Science* **85**, 630.

Meuller, J., Brun del Re, G., Buerki, H., Keller, H.U., Hess, M.W. and Cottier, H. (1975). Nonspecific acid esterase activity: a criterion for differentiation of T and B lymphocytes in mouse lymph nodes. *Eur. J. Immunol.* **5**, 270.

Moloney, W.C., McPherson, K. and Fliegelman, L. (1960). Esterase activity in leucocytes demonstrated by the use of naphthol AS-D chloroacetate substrate. *J. Histochem. Cytochem.* **8**, 200.

Nachlas, M.M. and Seligman, A.M. (1949). The histochemical demonstration of esterase. *J. Nat. Cancer Inst.* **9**, 415.

Okun, M.R., Edelstein, L., Nebaur, G. and Hamada, G. (1969). The histochemical tyrosine–DOPA reaction for tyrosinase and its use in localizing tyrosinase activity in mast cells. *J. Invest. Derm.* **53**, 39.

Pearse, A.G.E. (1953). *Histochemistry: theoretical and applied*, (1st edn). Churchill, London.

—— (1957). Intracellular localisation of deydrogenase systems using mono-tetrazolium salts and metal chelation of their formazans. *J. Histochem. Cytochem.* **5**, 515.

—— (1960). *Histochemistry: theoretical and applied,* (2nd edn). Churchill, London.

—— (1963). Rapid freeze drying of biological tissues with a thermoelectric unit. *J. Sci. Instr.* **40**, 176.

—— (1968). *Histochemistry: theoretical and applied* (3rd edn), Vol. 1. Churchill, London.

—— (1972). *Histochemistry: theoretical and applied,* (3rd edn), Vol. 2. Churchill-Livingstone, Edinburgh.

—— (1980). *Histochemistry: theoretical and applied,* (4th edn), Vol. 1. Churchill-Livingstone, Edinburgh.

Reynolds, G.J. (1982). *Lymphoid tissue: a histological approach.* Wright, Bristol.

Robinson, J.M. and Karnovsky, M.J. (1983). Ultrastructural localisation of several phosphatases with cerium. *J. Histochem. Cytochem.* **31**, 1197.

Rosenberg, M., Bartl, P. and Lesko, J. (1960). Water-soluble methacrylate as an embedding medium for the preparation of ultrathin sections. *J. Ultrasturct. Res.* **4**, 298.

Ruddell, C.L. (1967). Embeding media for 1–2 micron sectioning. 2. Hydroxyethyl methacrylate combined with 2-butoxyethanol. *Stain Technol.* **42**, 253.

Sabatini, D.D., Bensch, K. and Barrnett, R.J. (1963). Cytochemistry and electron microscopy. The preservation of cellular ultrastructure and enzymatic activity by aldehyde fixation. *J. Cell. Biol.* **17**, 19.

Scarpelli, D.G., Hess, R. and Pearse, A.G.E. (1958). The cytochemical localisation of oxidative enzymes. 1. Diphosphopyridine nucleotide diaphorase and triphosphopyridine nucleotide diaphorase. *J. Biophys. Biochem. Cytol.* **4**, 747.

Schlake, W., Meyer, E.M. and Grundman, E. (1978). Histochemical identification of T and B areas in paraffin-embedded lymphoid tissue by demonstration of α-napthyl acetate esterase (ANAE) activity. *Path. Res. Pract.* **163**, 173.

Seligman, A.M., Chauncey, H.H. and Nachlas, M.M. (1951). Effect of formalin fixation on activity of five enzymes of rat liver. *Stain Technol.* **26**, 19.

Shnitka, T.K. and Seligman, A.M. (1971). Ultrastructural localisation of enzymes. *Ann. Rev. Biochem.* **40**, 375.

Simpson, W.L. (1941). Experimental analysis of Altman's technique of freeze drying. *Anat. Rec.* **80**, 329.

Sjöstrand, F.S. and Bernhard, W.B. (1976). The structure of mitochondrial membranes in frozen sections. *J. Ultrastruct. Res.* **56**, 233.

Slezak, J. and Geller, S.A. (1984). Cytochemical studies of myocardial adenylate cyclase after its activation and inhibition. *J. Histochem. Cytochem.* **32**, 105.

Stäubli, W. (1960). Nouvelle matière d'inclusion hydrosoluble pour la cytogie électronique, *Comptes Rendu Hebdomadaire des Séances del'Académie des Sciences* **250**, 1137.

Thompson, G. and Germaine, J. (1984). Histochemistry and immunocytochemistry of fixation labile moieties in resin embedded tissue. *J. Path.* **142**, A6.

Tsou, K.C., Cheng, C.S., Nachlas, M.M. and Seligman, A.M. (1956). Synthesis of some *p*-nitrophenyl substituted tetrazolium salts as electron acceptors for the demonstration of dehydrogenase. *J. Am. Chem. Soc.* **78**, 6139.

———, Goodwin, C.W., Seamond, B. and Lynn, D. (1968). Intracristal localization of succinic dehydrogenase activity with a new osmium-containing tetra-tetra-zolium salt. *J. Histochem. Cytochem.* **16**, 487.

Wachstein, M. and Wolf, G. (1958). The histochemical demonstration of esterase activity in human blood and bone marrow smears. *J. Histochem. Cytochem.* **6**, 457.

Weller, R.O. (1984). Muscle biopsy and the diagnosis of muscle disease. In *Recent advances in histopathology*, (eds P.P. Anthony and N.M. Macsween), No. 12, p. 259. Churchill-Livingstone, Edinburgh.

Wilkinson, J.H. (1976). *The principles and practice of diagnostic enzymology*. Edward Arnold, London.

Wright, D.H. and Isaacson, P.G. (1983). *Biopsy pathology of the lymphoreticular system*. Chapman and Hall, London.

Yam, L.T., Li, C.Y. and Crosby, W.H. (1971). Cytochemical identification of monocytes and granulocytes. *Am. J. Clin. Path.* **55**, 283.

Index

TOP LEFT

Acid phosphatase activity seen as a purple reaction product demonstrated by a substituted naphthol method in rat liver. Tissue prepared by the freeze drying technique and embedded in a low-melting-point paraffin wax. Fixation of the section was with formaldehyde vapour.

TOP RIGHT

Non-specific esterase activity in a human jejunal biopsy demonstrated by the indoxyl acetate method. The enzyme hydrolyses 4-chloro-5-bromo-indoxyl acetate which is then oxidized by potassium ferricyanide to a blue–green dye. Cryostat section prefixed in 4 per cent formol calcium at $4\,^{\circ}$C.

BOTTOM LEFT

Sucrase activity on the villi of human jejunum histochemically stained orange-red using 6-bromo-2-naphthyl-α-D-glucoside and hexazonium pararosaniline. Tissue fixed in 4 per cent neutral formol calcium, washed in 3 per cent sucrose in 0.2M cacodylate buffer pH 7.4, and embedded in glycol methacrylate resin all at $4\,^{\circ}$C; 3μm section. Counterstained with (chloroform washed) methyl green.

BOTTOM RIGHT

Myeloid cells in rat spleen histochemically stained red for chloroacetate esterase using naphthol AS-D chloroacetate and hexazonium pararosaniline. Tissue fixed in 4 per cent neutral formol calcium, washed in 3 per cent sucrose in 0.2M cacodylate buffer pH 7.4, and embedded in LR White resin all at $4\,^{\circ}$C; 2μm section. Counterstained with haematoxylin.